1982

To Dave,

With many thanks for
your help and encouragement.

Jane

VISUAL CELLS IN EVOLUTION

Visual Cells in Evolution

Editor

Jane A. Westfall, Ph.D.
Department of Anatomy and Physiology
Kansas State University
Manhattan, Kansas

Raven Press ■ New York

Raven Press, 1140 Avenue of the Americas, New York, New York 10036

Made in the United States of America

International Standard Book Number 0–89004–760–X
Library of Congress Catalog Number 80–6277

To Professor Richard M. Eakin, Professor Emeritus, Department of Zoology, University of California at Berkeley, in recognition of his stimulating contributions to the study of photoreceptors in the animal kingdom.

Preface

New analytical techniques and comparative studies of photoreceptors have led to a revival of interest in the evolution of visual cells. This volume covers the latest theories on photoreceptor origins and discusses state-of-the-art techniques used in the study of visual cells in various invertebrate and vertebrate systems. A wealth of ultrastructural information on photoreceptors has been correlated with membrane biochemistry and turnover, morphogenetic experimentation, physiological changes, and evolution. Through the contributions presented herein students and researchers are introduced to unresolved problems and current controversies as a stimulus to future investigation of visual systems.

This volume details the biochemical differences in invertebrate and vertebrate photoreceptive membranes, their cyclic changes and ultrastructure as well as genetic specificity. New research on development of specific neuronal properties and identification of neurotransmitter candidates is presented, along with an original hypothesis for evolution of photoreceptor synapses. For the first time, three controversial theories supporting mono-, di-, and polyphyletic origins for photoreceptors are presented together for scientific comparison.

The overall emphasis of this book is on the patterns of continuity and diversity in visual cells as they evolve over time. This book aims to demonstrate a recognizable thread of design in photoreceptors within the animal kingdom so that logical hypotheses for their probable evolution can be made.

Jane A. Westfall

Acknowledgments

I gratefully acknowledge the National Science Foundation for support of the symposium on *Visual Cells in Evolution* that led to this publication. I also thank the American Society of Zoologists for hosting the symposium at its annual meeting, December 27–30, 1980 at Seattle.

For success of the symposium, I express sincere appreciation to Mary Wiley, Business Manager of the American Society of Zoologists, the panel discussants (Drs. Richard Bensinger, Dean Bok, Ann H. Bunt, Gordon L. Fain, Steven K. Fisher, Colin O. Hermans, Matthew La Vail, James O'Donnell, and Robert Rodieck), speakers, and additional contributors to this book.

For editorial assistance and manuscript retypings I thank Drs. John C. Kinnamon and David E. Sims, and Ms. Tari Metzler of the Department of Anatomy and Physiology, Kansas State University.

Contents

Contributors

Robert E. Anderson: *Department of Ophthalmology, Baylor College of Medicine, Texas Medical Center, Houston, Texas 77030*

Lary D. Andrews: *Department of Ophthalmology, Baylor College of Medicine, Texas Medical Center, Houston, Texas 77030*

Richard M. Eakin: *Department of Zoology, University of California, Berkeley, California 94720*

Jeanne M. Frederick: *Department of Ophthalmology, Baylor College of Medicine, Texas Medical Center, Houston, Texas 77030*

Joe G. Hollyfield: *Department of Ophthalmology, Baylor College of Medicine, Texas Medical Center, Houston, Texas 77030*

Dominic Man-Kit Lam: *Department of Ophthalmology, Baylor College of Medicine, Texas Medical Center, Houston, Texas 77030*

Robert E. Marc: *Department of Ophthalmology, University of Texas, Health Science Center, Houston, Texas 77025*

Mary E. Rayborn: *Department of Ophthalmology, Baylor College of Medicine, Texas Medical Center, Houston, Texas 77030*

Luitfried v. Salvini-Plawen: *Zoologisches Institut, Universität Wien, Wien 1, Austria*

P. Vijay Sarthy: *Department of Ophthalmology, Baylor College of Medicine, Texas Medical Center, Houston, Texas 77030*

O. Trujillo-Cenóz: *Laboratory of Comparative Neuroanatomy, Biological Science Research Institute, Montevideo, Uruguay*

J. R. Vanfleteren: *Laboratorium voor Morfologie en Systematick, Institut voor Dierkunde, Rijksuniversiteit Gent, B–9000 Gent, Belgium*

Talbot H. Waterman: *Department of Biology, Yale University, New Haven, Connecticut 06520*

Jane A. Westfall: *Department of Anatomy and Physiology, Kansas State University, Manhattan, Kansas 66506*

Visual Cells in Evolution, edited by Jane A. Westfall,
Raven Press, New York © 1982.

Biochemistry of Retinal Photoreceptor Membranes in Vertebrates and Invertebrates

Robert E. Anderson and Lary D. Andrews

Cullen Eye Institute, Baylor College of Medicine, Houston, Texas 77030

We would like to thank Dr. Jane A. Westfall for organizing this symposium honoring Dr. Richard M. Eakin. In addition we acknowledge Dr. Eakin's many contributions to visual science which have provided the impetus for this occasion.

In this paper we will compare and contrast the biochemistry of photoreceptor membranes from the retinas of a number of vertebrate and invertebrate species. Such a comparison is justified on the grounds that structural and functional differences exist between these two groups. The vertebrate retina contains a modified cilium which is absent from many invertebrate retinas. In response to light, vertebrate (ciliary) photoreceptors hyperpolarize whereas the rhabdomeric photoreceptors of invertebrates depolarize. Despite structural and functional differences there is one unifying feature: both use 11-cis-retinaldehyde as a chromophore, bound covalently to a specific membrane protein, to absorb photons and initiate visual excitation.

A large part of the biochemical studies on photoreceptor membranes is from our laboratory. No attempt has been made to include all of the literature pertaining to vertebrate photoreceptors. We have tried to cover most of the chemical literature on invertebrate photoreceptors, which is miniscule compared to that available for vertebrates. It should be noted that almost all of the biochemical studies of vertebrate photoreceptor membranes have been directed toward rods rather than cones because of the ease with which pure rod outer segments may be isolated in relatively high yield and the preponderance of rods in retinas of the most convenient experimental animals, namely, frog, rat, and cow.

Although Dr. Waterman will discuss the comparative morphology of vertebrate and invertebrate photoreceptor membranes in the next paper, a brief overview of this subject will be presented which is pertinent to understanding the context and relevance of some of our chemical data. Before proceeding, however, we should point out that the comparative anatomy and physiology of vertebrate and invertebrate photoreceptor membranes have been discussed by Wolken (81, 82) and the chemistry reviewed by Abrahamson et al. (1).

COMPARATIVE MORPHOLOGY OF VERTEBRATE AND INVERTEBRATE RETINAS

The photoreceptor cells of the vertebrate retina are the rods and cones. These cells are highly polarized with one extremity (the outer

1

segment) consisting of stacks of membranes that contain a light absorbing photopigment, and the other end making synaptic contact with bipolar and horizontal cells (and in some instances electrical contact with each other). The outer segment and its narrow connection to the rest of the cell is a modified cilium, and for this reason vertebrate photoreceptors are said to be of the ciliary type.

The photoreceptors or retinula cells of many invertebrate retinas, with a few exceptions, concentrate photopigment molecules in the plasma membranes of closely packed microvilli that extend from one side of the cell. In compound eyes these cells are organized into units called ommatidia, containing upwards to eight individual retinula cells. The microvillar membranes of neighboring retinula cells in an ommatidium are intimately intermingled in a regular, repeating pattern to form the rhabdom. Thus many invertebrate photoreceptors are said to be of the rhabdomeric type. This high degree of association between photoreceptor membranes in adjacent cells is not observed in the vertebrate retina.

The morphology of the photoreceptor membranes in vertebrate and invertebrate retinas is quite different. Each rod outer segment (ROS) contains from several hundred to a couple thousand free floating disks that are stacked like coins within a cylinder defined by the plasma membrane of the ROS. All but the most basal disks appear not to contact each other or the surrounding plasma membrane. A disk may be from 1-12 μm wide, depending upon the species examined, and they are regularly spaced with a repeat distance of about 30 nm. Each disk consists of two membranes fused at the edge to form a flattened saccule. The long axis of each outer segment is nearly parallel to the direction of light striking that part of the retina, and thus the plane of each disk is perpendicular to the incoming light rays. Unlike the disks of the vertebrate ROS, the microvillar processes are actually special extensions of the plasma membrane of the retinula cell. These finger-like projections are several micrometers long and have a cross-sectional diameter of around 50-70 nm. Thus the topology of rhabdomeric microvilli is analogous to that of most vertebrate cone outer segments in which the disk membranes and the plasma membranes of the inner and outer segments are in continuity.

ISOLATION AND CRITERIA FOR PURITY OF PHOTORECEPTOR MEMBRANES OF VERTEBRATE AND INVERTEBRATE RETINAS

Vertebrates

The procedures for highly purified preparations of vertebrate photoreceptor membranes have been worked out in great detail in a number of laboratories (11, 62, 76, 77). Most procedures make use of the fact that these membranes are less dense than those of other cellular organelles, and can be readily separated from the heavier membranes by continuous or discontinuous sucrose gradient ultracentrifugation.

Several criteria have been established for judging the purity of these preparations. One is the ratio of the absorbance at 278 nm (protein) to the change in absorbance at 498 nm (rhodopsin) following light exposure of membranes prepared in darkness or under dim red light. In our laboratory values less than 3.0 are taken as an indication of a reasonably pure preparation. Others have reported lower values for ROS prepared by sucrose density centrifugations, but these lower values usually incorporate a correction for light scattering that lowers the A-278 nm value and thus the A-278/ΔA-498 ratio. All of our reported ratios are uncorrected.

Electron microscopy of rod outer segments prepared by gentle mechanical dissociation techniques can be used to identify some organelle contaminants. However, vigorous homogenization usually results in the production of vesicles that have lost most of their native ROS structure.

Another estimation of vertebrate ROS purity is made by polyacrylamide gel electrophoresis. As will be discussed in a following section, the protein composition of vertebrate ROS membranes is quite simple. The major integral protein is rhodopsin; the only other protein studied in any detail is the high molecular weight protein described by Papermaster et al. (61) that amounts to at most 1-3% of the material on the gel. In our experience, at least 95% of the Coomassie blue stain on polyacrylamide gels of detergent-solubilized, extensively washed, dark-adapted ROS membranes is due to opsin.

Invertebrates

The criteria for purity of rhabdomeric microvillar membranes are not so rigorous as those for vertebrate photoreceptor membranes. One reason for this is that fewer studies have been carried out on invertebrate photoreceptors, and a second is that the yield of purified photoreceptor membranes from invertebrate retinas is not nearly so great as that from vertebrates. Mason et al. (53) have reported the successful isolation of invertebrate microvilli by use of a continuous sucrose gradient. Our attempts with discontinuous gradients have resulted in the recovery of photoreceptor membranes (as judged by the presence of bleachable photopigment) at all of the gradient interfaces. The most successful procedure we have used in isolating microvillar membranes is repeated flotations with buffered 42% sucrose, alternating with washings in buffer (6, 14). After several of these sedimentation-flotation steps, membranes of reasonable purity have been obtained, as judged by using the criteria established for vertebrate photoreceptors. Our preparation of Limulus (14) and squid (6) photoreceptor membranes gave uncorrected A-278/ΔA-490 photopigment ratios of 3.0 and 3.6 respectively. These values compare acceptably with the A-280/ΔA-494 ratios of 3.2 obtained by Hagins (38) and 2.8 obtained by Nashima et al. (58) for their best preparations of membranes containing squid photopigment. All of these are greater than the value of 2.3 for this ratio claimed by Mason et al. (53) for squid microvillar membranes. Although it is not clear from their paper, this latter value probably was corrected for scatter at 278 nm.

To our knowledge workers in no laboratory have used an ultrastructural evaluation as an estimate of the purity of their preparation of purified fractionated rhabdomeric microvilli.

Polyacrylamide gel electrophoresis of invertebrate photoreceptor membranes has been carried out in several laboratories (6, 38, 58, 64). In none of these was the relative abundance of the photopigment as great as that observed for routine preparations of vertebrate rod outer segments. Whether this indicates that current preparations of invertebrate photoreceptor membranes are not so pure as those obtained from vertebrates, or that invertebrate photoreceptor membranes contain a number of integral proteins other than photopigment, still remains to be determined.

CHEMISTRY OF VERTEBRATE AND INVERTEBRATE PHOTORECEPTOR MEMBRANES

The "backbone" of a biological membrane is the lipid bilayer that forms a passive barrier to diffusion of ions and molecules, and results in compartmentalization of the cell. The lipid bilayer also provides a matrix in which the integral membrane proteins are embedded. Specific properties of membranes are usually controlled by specific integral proteins (or protein complexes) so that there are unique proteins in membranes for transport, receptor binding, light absorption, etc.

The function of many membrane proteins requires that they be part of a lipid bilayer, and removal from the bilayer often results in loss of properties observed in the native membrane. For example, vertebrate rhodopsin requires lipid for regeneration following photobleaching (21, 74, 89). The bilayer also provides a stable matrix that partially protects rhodopsin from chemical and thermal insults (21).

Although biological membranes are seldom preponderantly composed of a unique lipid (as compared to protein), each membrane has a distinctive lipid class and fatty acid composition. This is important in membrane function because both lipid class and fatty acid composition determine membrane fluidity (viscosity), which in turn can modulate metabolic processes that involve transport across or reactions within the bilayer. A comparison of the chemistry of vertebrate and invertebrate photoreceptor membranes is of interest because these membranes perform similar tasks (photon capture), but nevertheless differ in morphology and molecular architecture.

Protein Chemistry

The protein chemistry of vertebrate photoreceptor membranes is relatively simple, as mentioned in the preceding sections. Rhodopsin is the major integral protein, comprising most of the Coomassie blue stained material on polyacrylamide gels. The only other integral protein that has been studied in vertebrate retinas is a high molecular weight protein (240,000 daltons) shown by Papermaster et al. (63) to be situated in the incisures and margins of frog red rod outer segment disks. The molecular weight of vertebrate rhodopsin may be as low as 27,000-30,000 (41, 75), determined by amino acid analysis, and as high as 60,000 (80), determined by density ultracentrifugation. These different values were usually derived from procedures for determination of the molecular weights other than SDS polyacrylamide gel electrophoresis. Basinger (personal communication) used this technique to examine the molecular weights of rhodopsin from a number of vertebrate species, ranging from frog to human, and found an average value of 36,000. With consideration for the inherent problems that may arise when using this technique to study membrane proteins (32), the concensus among most investigators is that the molecular weight of rhodopsin is about 36,000.

Careful analysis by Heller (42) of frog, cattle, and rat rhodopsin revealed that these molecules possess strikingly similar amino acid compositions. On the other hand, a recent report has appeared describing multiple forms of rhodopsin from frog ROS that differ in molecular weight by about 1,000 (56). Also, isoelectric focusing studies on cattle rhodopsin have revealed species of rhodopsin that have different pI values (65). Therefore, whereas vertebrate rhodopsins share many common features some differences may exist among and within the various species.

Even though the integral protein composition of vertebrate rod outer segments appears to be quite simple, a number of other proteins have been found to be associated with isolated ROS through their enzymatic

activities. These proteins may be either soluble or loosely bound to
the membranes. Kuhn (49) has observed that exposure of dark-adapted
cattle ROS to light results in the reversible binding of at least four
proteins with molecular weights of 68,000, 48,000, 37,000 and 35,000
daltons, as estimated by gel electrophoresis. The 68,000 protein was
shown to contain rhodopsin kinase activity, whereas the 35,000 and 37,000
molecular weight proteins were together shown to have GTPase activity
when bound to ROS membranes. At the present time, the function of the
48,000 protein is not known. Presumably the observed specific bindings
allow the control of these enzymatic activities by light. A large volume
of literature, which will not be reviewed in this paper, has shown that
vertebrate ROS contain enzymes that metabolize cyclic nucleotides, and
that these enzymes are modulated by light (see Lolly, 51, for review).
Other ongoing biochemical processes include protein phosphorylation and
energy metabolism, which also appear to be controlled at least in part
by light.

The most thoroughly studied invertebrate photoreceptor membranes are
those from the squid retina. Our laboratory (6), Nashima et al. (58),
and Hagins (38) reported values of 50,500, 51,000 and 49,000 daltons
respectively for the molecular weight of squid photopigment determined
by polyacrylamide gel electrophoresis. A value of 43,000 daltons was
obtained by Paulson and Schwimmer (64) for photopigment from the cephalopod Eledone moschata (octopus).

The amino acid compositions determined for the squid photopigment by
Hagins (38), Nashima et al. (58) and Abrahamson and Wiensenfeld (2)
clearly show more variability within this species than Heller found
among the rat, bovine, and frog (42). These differences probably are
due to the use of impure preparations of the invertebrate photoreceptor
membranes and the ensuing difficulties in obtaining enough pure protein
for accurate amino acid analysis. Also, as pointed out by Nashima
et al. (58), there is a rapid proteolysis of the squid photopigment
that leads to loss of a C-terminal peptide which is enriched in certain
amino acids. The data of these authors probably reflects the true amino
acid composition of the squid photopigment.

Rhodopsin in vertebrate rod outer segments is asymmetrically distributed in the lipid bilayer (69) with the carbohydrate moiety present on
the inside of the intact disk. Although no complete study of this type
has been published on invertebrates, reference to unpublished results
was made by Nashima et al. (58). They found that vesicles prepared from
squid photoreceptor membranes were not aggregated by concanavalin A
unless they were first disrupted by a detergent; this is consistent with
the hypothesis that the protein is asymmetrically distributed with respect to its carbohydrate moiety.

Lipid Chemistry

The lipid class compositions of a number of vertebrate and invertebrate photoreceptor membranes are given in Table I. In the vertebrates,
with the exception of goldfish, the phospholipids comprise about 85-90%
of the total lipids of the rod outer segments. The major two components
are phosphatidylethanolamine and phosphatidylcholine, but relatively
large amounts of phosphatidylserine are also found. Small but consistent amounts of sphingomyelin and phosphatidylinositol are observed in
these membranes as well. The major neutral lipid in vertebrate photoreceptor membranes is cholesterol, which is present as about 7-9 mole%

TABLE 1

Lipid Class Composition of Some Vertebrate and Invertebrate Species

Lipid Class	Human[a]	Rat[a]	Frog[a]	Goldfish[b]	Octopus[c]	Squid[d]	Limulus[e]	Blow-fly[f]	Moth[f]
	(Relative mole %)								
Glycerolipids (GL)									
Phospholipids									
Phosphatidyl choline	32.1	27.5	38.1	34.0	27.9	32.2	32.9	26.2	40.1
Phosphatidyl ethanolamine	37.1	38.1	29.1	20.6	32.7	30.6	19.7	57.8	38.9
Phosphatidyl serine	12.0	12.8	10.8	6.1	2.7	4.0	7.8	6.1	6.9
Phosphatidyl inositol	-	0.9	1.9	1.1	-	-	3.9	4.3	1.4
Lysophosphatidyl choline	-	0.6	0.6	-	-	1.7	1.2	0.6	6.4
Lysophosphatidyl ethanolamine	2.0	1.3	0.4	-	-	3.9	1.8	2.4	1.8
Diphosphatidyl glycerol	-	-	-	-	-	-	-	1.4	2.7
Unidentified	1.0	5.7	2.3	3.3	3.2	0.5	4.2	1.2	1.8
Neutral Lipids									
1,2-diglycerides (DG)	0.9	0.6	1.6	1.3	-	-	N.D.	N.D.	N.D.
Triglycerides	-	-	-	-	1.3	-	N.D.	N.D.	N.D.
Sphingolipids (SL)									
Sphingomyelin	1.8	0.7	1.6	3.1	1.8	1.3	11.4	-	-
Ceramide 2-aminoethyl phosphonate	-	-	-	-	3.7	-	-	-	-
Free Fatty Acids (FFA)	2.1	3.6	5.6	21.0	9.6	8.6	N.D.	N.D.	N.D.
Cholesterol (Chol.)	9.8	7.8	8.0	9.4	17.1	17.2	17.0	N.D.	N.D.
Total Neutral Lipids (DG+FFA+Chol+TG)	12.9	12.0	15.2	31.7	28.0	25.8	17.0	N.D.	N.D.
Total Phospholipids (GL+SL)	87.0	87.8	84.8	68.2	72.0	74.2	82.9	100.0	100.0

N.D. = Not Determined

TABLE 1 (footnote)

[a]Wiegand, R. D. and Anderson, R. E. (unpublished results).
[b]Fliesler, S. J., Maude, M. B. and Anderson, R. E. (unpublished results.
[c]Recalculation from the results of Akino, and Tsuda (4).
[d]Anderson et al. (11).
[e]Benolken et al. (14).
[f]Weber and Zinkler (79).

---

of the total lipid. This value is less than that usually found for cholesterol in biological membranes that are derivatives of plasma membranes. The other neutral lipids are 1,2-diglycerides and free fatty acids, which are minor constituents.

The lipid class compositions of invertebrate photoreceptor membranes are somewhat different from those of the vertebrates. The major phospholipids are still phosphatidylethanolamine and phosphatidylcholine, but the levels of phosphatidylserine are reduced. The neutral lipid content of the squid photoreceptor membranes appears to be higher than that of the vertebrates (except for the goldfish), and, as found in the vertebrates, the major component is cholesterol. Low levels of triglycerides were reported by Akino and Tsuda (4) for octopus photoreceptors. Cholesterol, cholesterol esters, and triglycerides were found in significant amounts by Mason et al. (53) in squid photoreceptors. However, our high temperature gas liquid chromatographic analysis of the neutral lipids of squid photoreceptors did not reveal the presence of cholesterol esters or triglycerides (6). Because this procedure is able to detect small amounts of these lipids (<1% of the total neutral lipid), our analysis indicates that these two lipid classes are not present in squid photoreceptor membranes.

Although there have been some reports of glycolipid and gangliosides in vertebrate (31, 43) and of glycolipids in invertebrate (1) photoreceptor membranes, we have found no evidence for these two lipid classes in any of the purified ROS membranes we have assayed. Akino and Tsuda (4) did not report any glycolipids in octopus photoreceptor membranes. Because the unfractionated retina is relatively rich in these lipids (31, 45), it is our impression that the glycolipids and gangliosides reported for photoreceptor membranes are contaminants from other retinal organelles.

Whereas some differences exist between the lipid class compositions of vertebrate and invertebrate photoreceptor membranes, they are relatively minor; the major differences are in their respective fatty acid compositions (Tables 2-4). The fatty acid nomenclature is as follows: the first number specifies the number of carbon atoms of the fatty acid, the number after the colon specifies the number of double bonds, and the number after ω specifies the position of the first double bond from the terminal methyl group. For example, 22:6ω3 is a 22 carbon fatty acid with 6 double bonds the first one situated between the third and fourth carbons from the methyl terminal. Two families of polyunsaturates are found in photoreceptor membranes; these are identified as ω3 and ω6 acids. ω6 fatty acids are derived from linoleic acid (18:2ω6) and ω3 acids are derived from linolenic acid (18:3ω3). Both families are "essential" dietary items, because neither can be synthesized de novo by vertebrates or invertebrates. Tables 2, 3, and 4 contain data regarding the lipid composition of photoreceptor membranes from a variety of vertebrate and invertebrate retinas. Looking first at the vertebrates, the major polyunsaturate in all classes is 22:6ω3. Small amounts of 20:4ω6 and 20:5ω3 are present. The major saturated fatty acid in phosphatidylethanolamine and phosphatidylserine is stearic acid (18:0). Palmitic acid (16:0) is the major saturate of phosphatidylcholine. This composition is fairly uniform among all of the vertebrates we have examined.

Less similarity is found among the fatty acids of invertebrate photoreceptor membrane phospholipids. In <u>Limulus</u> and the moth, 20:5ω3 is the major fatty acid, and little if any 22:6ω3 is found. On the basis of these earlier observations, we thought perhaps that the absence of 22:6ω3 may be a unique feature of invertebrate retinas, especially because Mason et al. (53) had reported very low levels (less than 10%) of 22:6ω3 in the total fatty acids from squid photoreceptors. However, more detailed analysis in our laboratory (6) and by Akino and Tsuda (4) clearly established that cephalopod photoreceptor membranes contain large amounts of 20:4ω6, 20:5ω3, and 22:6ω3. In fact, cephalopod photoreceptor membranes contain the highest levels of polyunsaturated fatty acids of any photoreceptor membrane that has been analyzed to date.

Even though the levels of long chain polyunsaturates are much higher in squid and octopus photoreceptors than any other, vertebrate or invertebrate, we must emphasize that <u>all</u> photoreceptor membranes contain large amounts of long chain polyunsaturated fatty acids. There have been no other membrane systems shown to contain higher levels of 22:6ω3.

It has long been known that the fatty acids of phospholipids are distributed with the saturated fatty acids predominantly occupying the 1-position of the glycerol molecule and the polyunsaturated fatty acids occupying the 2-position. However, in those phospholipid classes where polyunsaturates comprise more than 50% of the fatty acids, clearly some polyunsaturates must be localized in the 1-position. A recent analysis of the molecular species of frog photoreceptors in our laboratory by Dr. Rex Weigand (unpublished) revealed the presence of large amounts (60 mole %) of dipolyunsaturated species of phosphatidylethanolamine. Aveldano de Caldironi and Bazan (13) and Miljanich et al. (55) have reported similar findings in bovine outer segments. An interesting observation from our studies was that 22:6ω3 is the preferred polyunsaturate for the 2-position, and the other polyunsaturates were almost exclusively situated at the 1-position. Clearly in the case of the cephalopod phospholipids, which contain more than 50% polyunsaturates, dipolyunsaturated fatty acid species must also be present. We did not determine the molecular species distribution of phospholipids in our squid photoreceptor membrane preparations; however, Akino and Tsuda (4) carefully analyzed the molecular species in the octopus phospholipids. In phosphatidylethanolamine, the longer chain polyunsaturated fatty acids were preferentially esterified to the 2-position, whereas shorter chains were found at the 1-position. Thus, the order preference appears to be a common feature of vertebrates and invertebrates. Likewise, most of the dipolyunsaturated species in squid were found in phosphatidylethanolamine, with only a small amoung (3.6 mole %) present in phosphatidylcholine. This is essentially what we observed for the distribution of molecular species in these phospholipid classes isolated from frog photoreceptor membranes.

The presence of large amounts of dipolyunsaturated phosphoglycerides appears to be a unique feature of photoreceptor membranes. These molecules provide a fluidity to the lipid bilayer in which the visual pigment proteins reside. One question that will be addressed below is

TABLE 2

Fatty Acid Composition of Phosphatidyl Ethanolamine From Some
Vertebrate and Invertebrate Species

Fatty Acid[a]	Human[b]	Rat[c]	Frog[b]	Goldfish[b] (Relative mole %)	Octopus[b]	Squid[b&c]	Limulus[b]	Moth[d]
14:0	0.1	0.2	0.2	0.3	0.5	0.3	-	-
15:0	-	-	0.1	0.2	-	-	-	-
16:0 DMA[e]	0.4	0.4	0.6	1.1	-	-	1.4	-
16:0	9.7	5.5	6.4	6.4	6.1	3.6	2.2	4.1
16:1ω7	1.0	0.3	1.1	2.5	-	0.1	0.9	0.3
17:0	-	-	-	0.4	0.4	0.4	0.5	-
18:0 DMA	0.5	1.2	0.9	0.7	0.5	-	12.0	-
18:0	36.2	28.1	13.0	13.4	1.8	3.7	11.7	9.5
18:1ω9	6.2	3.2	6.0	8.1	4.5	0.9	4.3	20.8
18:2ω6	0.3	0.6	0.8	0.9	0.3	-	1.8	4.7
18:3ω3	-	-	-	-	-	-	-	18.9
20:0	0.8	-	-	0.2	-	-	-	-
20:1	0.6	0.8	-	1.1	3.3	0.6	11.6	-
20:2ω6	-	-	-	-	-	-	-	-
20:3ω9	1.3	-	0.5	0.3	-	0.2	-	-
20:4ω6	3.5	2.9	6.5	2.4	27.7	34.0	12.1	1.4
20:5ω3	-	0.1	-	0.5	13.4	7.5	40.5	40.0
22:4ω6	1.2	0.2	7.0	0.4	-	-	-	-
22:5ω6	2.0	1.1	2.2	0.7	-	-	-	-
22:5ω3	0.8	-	3.2	1.2	-	0.1	-	-
22:6ω3	34.2	54.8	51.1	58.0	41.7	47.3	-	-
24: polys.	0.1	-	-	0.8	-	-	-	-

TABLE 2 (footnote)

[a]Nomenclature defined in footnote in the text.
[b]References given in TABLE 1.
[c]Anderson et al. (7).
[d]Zinkler (88).
[e]DMAs are dimethyl acetals derived from the acid methanolysis of plasmalogens.

---

an apparently paradoxical difference between vertebrate and invertebrate photopigments: rhodopsin is free to undergo both rotational (22, 27) and translational (67) diffusion within the plane of the bilayer of vertebrate photoreceptor membranes, whereas the visual pigments of crayfish (37) and barnacle (5) microvilli appear to be relatively immobile in the bilayer.

RENEWAL OF PHOTORECEPTOR MEMBRANES

Vertebrates

Over a decade ago it was discovered that the photoreceptor membranes of vertebrate retinas were constantly being renewed. In a series of elegant autoradiographic studies (see Young, 86 and 87 for review) it was shown that protein was synthesized in the inner segment, transported to the base of the outer segment, and incorporated into forming disks. In rods, these eventually pinch off and the free floating disks thus formed displace older disks in an apical direction. The oldest disks at the tip of the ROS are shed as discrete membrane packets and phagocytized by the retinal pigment epithelium. A tight band of silver grains was observed throughout the apical migration, suggesting that throughout its lifetime in the ROS, the labeled protein remained within the disks into which it was originally incorporated. These autoradiographic studies were soon combined with biochemical studies (40) which showed that almost all of the labeled protein was rhodopsin, and that the specific radioactivity of rhodopsin in the ROS remained constant as long as the radioactive band was visible. Once the radioactive band reached the tip of the ROS and was shed, there was a precipitous drop in the specific radioactivity of rhodopsin in isolated ROS. This elegant combination of biochemistry and autoradiography convincingly demonstrated that rhodopsin, once incorporated into a growing ROS, remained with that disk throughout its duration in the ROS.

Recent studies from the laboratories of Papermaster and of Besharse have shed new light on the mechanism of transfer of newly synthesized protein destined for incorporation into growing ROS. Using immunoelectrophoresis techniques to study the early time course of opsin synthesis in the frog retina, Papermaster et al. (63) observed that the protein was membrane bound from the earliest time at which it could be detected. Thus, opsin must be synthesized in association with membranes and must remain in a membrane until incorporated into the growing ROS. Since there is some distance between the site of synthesis in the inner segment and the point of incorporation into the basal infoldings of the ROS, a mechanism must exist for transporting newly synthesized opsin through the inner segment. Papermaster et al. (60) have suggested that this transport is mediated via vesicles that accumulate in the periciliary cytoplasm at the apex of the rod inner segment just beneath the plasma

TABLE 3

Fatty Acid Composition of Phosphatidyl Choline From Vertebrate and Invertebrate Species[a]

Fatty Acid	Human	Rat	Frog	Goldfish	Octopus	Squid	Limulus	Moth
				(Relative mole %)				
14:0	0.8	0.9	1.6	1.1	0.4	0.7	–	–
15:0	–	0.3	0.2	0.3	–	0.7	–	–
16:0 DMA	–	–	–	0.2	–	–	–	–
16:0	32.5	29.5	33.4	29.7	36.0	32.7	9.8	12.5
16:1ω7	3.9	1.0	6.5	5.0	–	0.3	3.7	0.7
17:0	–	0.7	0.2	0.5	0.5	0.4	1.3	–
18:0 DMA	0.1	–	–	0.3	–	–	–	–
18:0	16.8	11.7	21.9	14.6	1.8	1.4	6.3	26.3
18:1ω7	14.9	8.3	8.7	17.7	5.1	3.9	29.5	14.9
18:2ω6	1.1	0.8	0.8	1.2	1.4	0.6	2.7	6.2
18:3ω3	–	–	–	–	–	–	–	28.9
20:0	0.9	–	T	0.1	–	–	–	–
20:1	0.6	0.8	–	1.1	1.1	1.0	–	–
20:2ω6	–	–	T	–	–	–	–	–
20:3ω9	1.2	–	–	0.3	–	0.7	–	–
20:4ω6	4.7	3.4	1.7	1.2	14.9	17.3	14.0	2.2
20:5ω3	–	–	–	0.4	7.6	4.6	27.6	8.1
22:4ω6	0.4	–	0.4	0.5	–	–	–	–
22:5ω6	0.9	3.2	0.7	0.7	–	0.1	0.6	–
22:5ω3	0.6	–	0.4	0.4	–	0.1	–	–
22:6ω3	19.5	38.4	23.6	24.4	30.1	34.3	0.8	–
24: polys	–	0.4	–	0.5	–	–	–	–

[a]See Tables 1 and 2 for references.

TABLE 4

Fatty Acid Composition of Phosphatidyl Serine From Some
Vertebrate and Invertebrate Species[a]

Fatty Acid	Human	Rat	Frog	Goldfish	Squid	Limulus	Moth
				(Relative mole %)			
14:0	0.5	-	1.3	0.9	0.9	-	-
15:0	0.1	-	1.5	0.2	0.4	-	-
16:0 DMA	-	-	-	0.2	-	1.1	-
16:0	2.6	2.5	4.0	9.6	10.1	3.1	4.2
16:1ω7	0.5	-	2.4	2.2	1.4	0.8	0.8
17:0	-	-	-	1.0	1.5	1.2	-
18:0 DMA	0.1	-	-	0.2	-	-	-
18:0	28.1	29.1	19.3	27.5	19.0	37.5	34.3
18:1ω9	5.9	1.9	3.4	5.1	1.7	22.8	16.2
18:2ω6	0.4	0.7	0.6	1.3	0.3	5.6	3.4
18:3ω3	-	-	-	-	-	-	12.3
20:0	1.2	-	0.1	0.4	-	0.1	-
20:1	1.2	-	-	1.0	3.8	-	-
20:2ω6	0.1	0.8	-	-	-	-	-
20:3ω9	0.8	-	-	0.4	-	-	-
20:4ω6	1.6	1.9	2.2	1.3	17.3	15.1	1.9
20:5ω3	-	-	-	0.2	6.2	7.2	25.4
22:4ω6	6.8	-	6.8	2.6	0.5	0.8	-
22:5ω6	2.2	-	4.5	1.0	-	0.4	-
22:5ω3	2.7	-	3.1	1.7	0.6	0.2	-
22:6ω3	34.1	61.8	46.4	42.3	33.0	-	-
24: polys	10.1	-	4.0	0.9	-	-	-

[a]See Tables 1 and 2 for references.

membrane. Such vesicles were described by Kinney and Fisher (46) in an earlier paper. Besharse and Pfenninger (15) used freeze-fracture electron microscopy to study these vesicles and found that their intramembranous particles were of the same dimensions as those present in the photoreceptor disks. Also, Papermaster and Besharse (59) have shown by immunocytochemical techniques that these vesicles contain opsin, and that they become labeled soon after exposure to radioactive protein precursors.

Because vesicles are not found within the connecting cilium, transfer of newly synthesized opsin to forming disks is probably accomplished by fusion of the vesicles with the inner segment plasma membrane. Support for this model has been recently obtained. In freeze-fracture replicas from retinas of frogs and goldfish, pairs of parallel rows of intramembrane particles were seen radially oriented around the base of the connecting cilium in the plasma membrane at the apex of the inner segment (12). These particle rows were situated atop shallow ridges in the membrane, and "dimples" resembling sites of vesicular exo- or endocytosis were occasionally found adjacent to them. These structures are strikingly similar to membrane specializations previously described in the neural terminal of the frog neuromuscular junction (44), and it has been observed that fusion events in this membrane occur in close association with these structures.

Although few studies have specifically dealt with the renewal of cone outer segments (23, 71), the available data indicate that the synthetic events parallel those in the rods. However, once the cone visual pigment has been incorporated into the outer segment plasma membrane, it freely diffuses throughout the entire outer segment, resulting in a diffuse autoradiographic labeling pattern. This is possible because all of the cone outer segment disk membranes are continuous, in contrast to the freefloating disks of the rods.

We have recently studied lipid renewal in vertebrate ROS (8-11). Lipids synthesized on the microsomes of the inner segment are incorporated into the growing ROS, and it seems reasonable to assume that both newly synthesized protein and lipid are transported by common vesicles to the site of insertion into the basal infoldings of the ROS plasma membrane. However, once incorporated into the ROS, the fate of the lipid is different from that of the protein. Our autoradiographic studies (9) confirm earlier ones of Bibb and Young (16, 17) that showed a diffuse labeling of lipids in the ROS following the injection of either labeled glycerol or fatty acids. Our biochemical studies on injection into the dorsal lymph sac of a precursor that is rapidly catabolized (such as glycerol or serine) clearly show that maximum specific activity of the phospholipids in the ROS is achieved early, but unlike protein, the specific activity immediately declines in an exponential manner. This indicates that the lipid, once incorporated into a basal disk of the ROS, is not confined to that disk throughout its apical displacement, but rather is free to diffuse throughout the entire ROS. Calculations of the half-life of phospholipid molecules in ROS gave values of 18-19 days. Based on the known turnover time of opsin in these frog ROS (39 days at a body temperature of 25°C), we calculated that the lipid was turning over in photoreceptor membranes more rapidly than could be explained simply on the basis of loss through shedding of tips.

Invertebrates

Photoreceptor membrane renewal in invertebrates displays considerable

variety. Membrane breakdown and reassembly have been observed morphologically in several species, and an interesting variability was evident
in the fate of the broken down membrane. Shedding into extracellular
space with subsequent uptake either by nonpigmented glial cells (20) or
by the photoreceptors (80) was observed or else shedding did not occur
but was replaced by endocytosis and autophagy (19, 25). Very few
biochemical studies have been carried out on the renewal of the components of invertebrate photoreceptor membranes. Several years ago at the
Bermuda Biological Station for Research, we injected tritiated leucine
into the pericephalic sinus of the squid and followed the incorporation
of radioactivity in photoreceptor membrane proteins (R. E. Anderson and
M. B. Maude, unpublished results). We observed an increase in the specific radioactivity of the major protein (presumably the photopigment)
over the 3 1/2 days of the experiment, suggesting that the photopigment
protein is renewed in these membranes. The difficulties in maintaining
squid in captivity for long periods of time precluded our following the
turnover of radioactive proteins in these membranes. However, what is
apparent from these preliminary studies is that there is an active metabolism of protein in the squid retina. Similar interpretations were made
in autoradiographic (47) and biochemical (48) studies following the incorporation of labeled leucine into the lateral eyes of Limulus. These
authors found one major difference between the patterns of renewal of
proteins in vertebrate and in invertebrate photoreceptor membranes:
whereas the site of synthesis and the pathway of migration of newly synthesized proteins can easily be demonstrated in vertebrates by autoradiographic techniques, such is not the case in Limulus. Krauhs et al.
(48) attribute this to the free access of microvilli to the cytoplasm,
the rapid labeling, and the retention of free label within the cell. In
one respect, the labeling pattern in invertebrate microvilli resembles
that of cone outer segments that are also diffusely labeled following
administration of a radioactive protein precursor.

PHYSICAL CHEMISTRY OF VERTEBRATE AND INVERTEBRATE PHOTORECEPTOR MEMBRANES

Vertebrates

The physical chemistry of the ROS membranes of several vertebrate
retinas has been characterized by a variety of sophisticated techniques.
To summarize these findings briefly, we know from x-ray (18, 35) and
neutron diffraction (72, 85) studies that the membranes are in a bilayer
configuration, that the hydrophobic core of the integral membrane proteins (almost entirely rhodopsin, as discussed above) is apparently distributed symmetrically through the thickness of the bilayer, and that
these proteins are randomly arrayed in the membrane with an average
intermolecular spacing of about 5 nm. Whether protein mass is symmetrically distributed between the head group regions on the two sides of
the disk membrane remains unresolved (see 35 and 85). From microspectro-
photometric studies we know that rhodopsin is free both to rotate (22, 27)
and to move laterally (66, 67) at a rapid rate within the plane of the
membrane at physiological temperatures. We also know from the dichroism
of outer segments that rhodopsin is not free to tumble about an axis
within the plane of the membrane that allows the unbleached chromophore,
11-cis-retinal, to maintain an optimal orientation for photon capture
(73).

Biochemical studies involving lectin binding to the oligosaccharide moiety of rhodopsin have demonstrated that this part of the molecule is present exclusively on the lumenal surface of the disk mambrane (3, 26). Other studies have shown that rhodopsin is sensitive to proteolysis acting only on the cytoplasmic side of the membrane (27, 33). Together these observations indicate that rhodopsin is a transmembrane protein. This is further supported by the finding that in reconstituted systems, in which the orientation of rhodopsin has been randomized, all rhodopsins are iodinated by the application of lactoperoxidase to one side of the membrane (34).

The overall shape and dimensions of the rhodopsin molecule have been studied by several biophysical techniques. The hydrophobic core of rhodopsin contains $\propto$-helices oriented perpendicular to the plane of the disk membrane (24, 54, 70). Fluorescence energy transfer experiments have been reported (83) in which various exogenous fluorescence chromophores were covalently linked to rhodopsin at distinct sites. In that study, the distances between several exogenous chromophores were calculated, as were the distances between each of these probe molecules and the endogenous 11-cis-retinal. It was concluded that rhodopsin has a length of at least 7.5 nm. With a molecular weight of about 40,000 daltons (their value), which would produce a spherical molecule with a radius of only about 2 nm, rhodopsin is thus indicated to have an elongated shape easily capable of spanning the disk membrane as was indicated to be the case by other studies described above.

In drawing correlations between structure, function, and composition it is important to consider the possible non-uniformity of the samples from which the data describing these membrane properties are obtained. In particular, compositional heterogeneity may exist both laterally within the plane of the membrane and between the two sides of the membrane. This statement is true for membrane proteins, lipids, and sugars. The distribution of rhodopsin both across the thickness of the disk membrane and laterally within the membrane has been described above. As also pointed out above, the sugars attached to rhodopsin are probably found exclusively on the lumenal surface of the disk membrane. The distribution of the various lipid components has proven to be complex and interesting. Several studies have found that phosphatidylethanolamine is asymmetrically distributed between the inner and outer leaflets of the disk membrane, with about 70% being present in the outer monolayer (28, 50,68). Contrasting results suggesting a nearly symmetrical transbilayer lipid distribution have also been reported (30), and thus this issue remains controversial. It has also been proposed that, relative to phosphatidylethanolamine, phosphatidylserine is preferentially found to be associated with rhodopsin in bovine ROS membranes (28, 29, 78). All of this indicates that the effective properties of the disk membrane at some locations may differ substantially from the average properties. An especially intriguing suggestion is that, since diskal phosphatidylcholine contains a significant amount of disaturated species (unpublished results from our laboratory), a transbilayer asymmetry in diskal lipids may result in the inner leaflet of the disk membrane (facing its lumen) being less fluid than the outer leaflet. The functional consequences of such a "transmembrane fluidity gradient" are unknown, but it is known that rhodopsin retains its spectral integrity and regenerability upon extraction from native disk membranes and reconstitution into symmetrical artificial membranes (21), indicating that an asymmetric membrane is not important for these functions.

Invertebrates

In contrast to the rather detailed picture of ROS membranes in the vertebrate retina, very little is known of the precise molecular architecture of microvillar membranes from invertebrate retinular cells. The sensitivity to polarized light described in many invertebrate species has long been associated with the possibility that chromophores in rhabdomeric microvillar membranes are not randomly oriented. It has been hypothesized that the chromophore axis is not only nearly parallel to the membrane surface, as is the case in vertebrate disks, but also has a tendency to be aligned parallel to the long axis of the microvilli (82). Such an orientation is inconsistent with a free rotational mobility such as that observed for vertebrate rhodopsin in disk membranes. It has been confirmed directly in the crayfish photoreceptor that both rotational and translational motions of the photopigment are restricted (37). Restricted translational mobility has also been observed in unfixed barnacle rhabdomeres (5). Yamada (84) has reported freeze fracture observations of rhabdomeric microvilli featuring molecular organization at an even higher level, in that the intramembrane particles were often arranged in rows.

To explain the restricted mobility of these invertebrate photopigments in situ, it has been speculated that one factor which may be involved is the lipid composition of the membrane (36). This idea was based upon the data of Mason et al. (53), but as discussed above, more current results do not support this suggestion. The polyunsaturated fatty acid content has been found to be high, and the cholesterol content, although greater than that in vertebrate ROS disks, is certainly insufficient by itself grossly to restrict protein mobility. A comparison of phase transition temperature of squid and frog photoreceptor membranes determined by differential scanning calorimetry revealed no differences (52).

Another factor that might restrict the mobility of invertebrate photopigments is a high protein:lipid ratio in the photoreceptor membranes. We have reported a ratio of 3:1 (by weight) for squid photoreceptor membranes (6), whereas Akino and Tsuda (4) found a value close to 1:1 for the octopus. The source of this apparent discrepancy is unknown. Determination of the protein:lipid ratio in crayfish photoreceptor membranes, for which the best data regarding photopigment immobility data are available, would clearly be of great interest.

Goldsmith (36) has summarized various reports on dichroic properties of rhabdomeres and presents an intriguing correlation between such properties and the ionic content of the bathing media. In particular, membranes maintained in media containing 2.6 mM Mg^{++} and 13.6 mM Ca^{++} had lower dichroic ratios than those in media with 28 mM Mg^{++} and 2.9 mM Ca^{++}. These data, however, do not distinguish between chromophore alignment and mobility. One obvious difference between ROS disks and rhabdomeric microvilli is the curvature of the membranes of the latter. It is possible that, if the hydrophobic part of the photopigment is elongated within the plane of the membrane, there might be a small tendency for the molecule to be aligned with the long axis of the microvilli.

CONCLUSION

Whereas it is clear that numerous hypotheses may be generated to integrate the facts presented above, it is evident that what is most needed is an expanded data base. Verification of Goldsmith's results in the

squid and octopus would be desirable, as would quantification of the
crayfish rhabdomeric lipid composition. There is one report (39) that
the dichroic ratio is 6 in the squid. The physical state of rhabdomeric
microvillar membranes and even the basic conformation of the lipids are
unknown. Reconstitution of invertebrate photopigments into vesicles
formed of native or other lipids could be performed to study how lipid-
lipid, lipid-protein, and protein-protein interactions might result in
the restricted mobility of invertebrate opsin in the native membranes.

At the beginning of this paper we proposed to use biochemical information in an attempt to analyze some of the structural and functional
differences found between vertebrate and invertebrate photoreceptors.
One of the most obvious differences is that the former hyperpolarize at
light onset whereas the latter depolarize. A major similarity is that.
in both, the membrane potential is modulated as a function of the membrane's sodium current. The biochemical information we have presented
cannot yet be directly linked to these functional properties of the
photoreceptor membranes, but it may be hypothesized that membrane proteins, including sodium channels, will eventually be isolated and characterized chemically, and the influence on them by their lipid microenviron-
ment elucidated.

Another function that both types of photoreceptor have in common is
the renewal of their photopigment-bearing membranes, although the detailed mechanisms differ. This topic has been briefly discussed above.
To recapitulate: the distal tips of photoreceptor outer segments in the
vertebrate retina are shed by the cell and subsequently phagocytized by
the adjacent cells of the pigment epithelium; in contrast, renewal of
rhabdomeric membranes has been shown in a number of cases to occur either
by direct internalization of the membrane or by shedding into extracellular space with subsequent pinocytosis by the photoreceptor. This
functional distinction may perhaps be correlated with the contrast between the narrow ciliary connection between the photoreceptive membranes
and the bulk of the metabolic machinery of the vertebrate cell, and the
broader cellular continuity between these compartments in the invertebrate rhabdomere. The processing of ingested membrane in the latter may
require a larger connection between the rhabdomeric microvilli and the
rest of the cell. Current biochemical data on both the vertebrate and
invertebrate membranes support the hypothesis that their lifelong renewal is necessitated by degradative processes associated with the high
degree of unsaturation in their constituent phospholipids and their
necessary exposure to oxygen and light. Exploration of this hypothesis
in the vertebrate retina through a detailed study of the biochemistry
of membrane damage owing to such treatments as prolonged exposure to constant light is ongoing in our and other laboratories.

A fundamental difference between the disks of rods and cones, and
rhabdomeric microvilli is that membranes of the former are planar whereas
those of the latter are cylindrical. The function of the microvillar
arrangement may be correlated with the motional restriction and alignment
of photopigment molecules in these membranes and the role this may play
in the sensitivity to polarized light. The available biochemical data
strongly suggest that special intermolecular forces must underlie this
motional restriction, because the membrane lipids are apparently highly
fluid at physiological temperatures, and membrane curvature alone seems
quite unlikely to be responsible for the marked restriction of translational mobility.

This last point leads to a more general notion regarding the comparative structure and biochemistry of photoreceptors. Visual transduction and the modulation of membrane potentials are molecular events, the requirements of which tailor the biochemistry of the photoreceptor cell, including the composition of photoreceptive membranes. As pointed out above, the biochemical requirements for the renewal of labile components may influence photoreceptor cells with different cellular environments to adopt both different strategies for membrane replacement and different associated structures as well. The cylindrical structure of rhabdomeric microvilli may be an adaption required to achieve the survival advantage of polarized light sensitivity, rather than a secondary consequence of the structure of more primitive cells from which they were derived.

The above discussion is of course speculative. However, it does serve to illustrate the importance of a detailed knowledge of photoreceptor biochemistry in the overall understanding of photoreceptor structure, function, and evolution.

ACKNOWLEDGMENTS

It is a great pleasure to acknowledge the contributions of many collaborators over the past decade. Specifically, thanks are given to R. M. Benolken, M. B. Maude, R. D. Wiegand, P. A. Dudley, P. A. Kelleher, S. F. Basinger, S. J. Fleisler, N. M. Giusto, M. B. Jackson, W-C Hsieh, P. M. Lissandrello, and T. M. Maida. We also thank Dr. Wolfgang Sterrer and his associates at the Bermuda Biological Station, St. Georges West, Bermuda, for providing support and facilities for our studies on squid photoreceptors.

Our studies have been supported over the years by grants from The Retina Research Foundation (Houston), Research to Prevent Blindness, Inc., Fight for Sight, Inc., the National Science Foundation, the National Retinitis Pigmentosa Foundation, the National Institutes of Health (NEI), and the Brown Foundation (Houston).

REFERENCES

1. Abrahamson,E.W.,Fager,R.S., and Mason,W.T.(1974): *Exp. Eye Res.*, 18:51-67.

2. Abrahamson,E.W., and Wiensenfeld,J.(1972): In: *Handbook of Sensory Physiology*, edited by H. J. A. Dartnell, pp. 69-121. Springer-Verlag, New York.

3. Adams,A.J.,Tanaka,M., and Shichi,H.(1978): *Exp. Eye Res.*, 27:595-605.

4. Akino,T., and Tsuda,M.(1979): *Biochim. Biophys. Acta*, 556:61-71.

5. Almagor,E.,Hillman,P., and Minke,B.(1979): *Biophys. Struct. Mechanism*, 5:243-248.

6. Anderson,R.E.,Benolken,R.M.,Kelleher,P.A.,Maude,M.B., and Weigand,R.D.(1978): *Biochim. Biophys. Acta*, 510:316-326.

7. Anderson,R.E.,Benolken,R.M.,Jackson,M.B., and Maude,M.B.(1977): In: *Function and Biosynthesis of Lipids*, Vol. 83, edited by N.G. Bazan, R. R. Brenner, and N. M. Giusto, pp. 547-559. Plenum, New York.

8.   Anderson,R.E.,Kelleher,P.A., and Maude,M.B.(1980): <u>Biochim. Biophys.</u>
     <u>Acta</u>, 620:227-235.

9.   Anderson,R.E.,Kelleher,P.A.,Maude,M.B., and Maida,T.M.(1980):
     <u>Neurochemistry</u>, 1:29-42.

10. Anderson,R.E.,Maude,M.B., and Kelleher,P.A.(1980): <u>Biochim. Biophys.</u>
     <u>Acta</u>, 620:236-246.

11. Anderson,R.E.,Maude,M.B.,Kelleher,P.A.,Maida,T.M., and Basinger,
     S.F.(1980): <u>Biochim. Biophys. Acta</u>, 620:212-226.

12. Andrews,L.D.(1981): In: <u>Structure of the Eye IV. Proceedings of the</u>
     <u>Fourth International Symposium on the Structure of the Eye</u>, edited
     by J.G. Hollyfield and E.A. Vidrio, Elsevier/North-Holland, (in
     press).

13. Aveldano de Caldironi,M.I., and Bazan,N.G.(1977): In: <u>Function and</u>
     <u>Biosynthesis of Lipids</u>, Vol. 83, edited by N. G. Bazan, R. R.
     Brenner and N. M. Guisto, pp. 397-404. Plenum, New York.

14. Benolken,R.M.,Anderson,R.E., and Maude,M.B.(1975): <u>Biochim. Biophys.</u>
     <u>Acta</u>, 413:234-242.

15. Besharse,J., and Pfenninger,K.(1980): <u>J. Cell Biol.</u>, 87:451-463.

16. Bibb,C., and Young,R.W.(1975): <u>J. Cell Biol.</u>, 61:327-343.

17. Bibb,C., and Young,R.W.(1974): <u>J. Cell Biol.</u>, 62:378-389.

18. Blaurock,A.E.(1977): In: <u>Vertebrate Photoreception</u>, edited by
     H.B. Barlow and P. Fatt, pp. 61-76. Academic Press, New York.

19. Blest,A.D.,Kao,L., and Powell,K.(1978): <u>Cell Tissue Res.</u>, 195:425-444.

20. Blest,A.D., and Maples,J.(1979): <u>Proc. R. Soc. Lond. (Biol)</u>,
     204:105-112.

21. Bonting,S.L.,van Bruegel,P.J.G.M., and Daeman,F.J.M.(1977): In:
     <u>Function and Biosynthesis of Lipids</u>, Vol. 83, edited by N. G. Bazan,
     R. R. Brenner, and N. M. Giusto, pp. 175-189. Plenum, New York.

22. Brown,P.K.(1972): <u>Nat. New Biol.</u>, 236:35-38.

23. Bunt,A.N.(1978): <u>Invest. Opthalmol. Vis. Sci.</u>, 17:90-104.

24. Chabre,M.(1978): <u>Proc. Natl. Acad. Sci. USA</u>, 75:5471-5474.

25. Chamberlain,S.C., and Barlow,R.B.,Jr.(1979): <u>Science</u>, 206:361-363.

26. Clark,S.P., and Molday,R.S.(1979):<u>Biochemistry</u>, 18:5868-5873.

27. Cone,R.A.(1972): <u>Nat. New Biol.</u>, 236:39-43.

28. Crain,R.C.,Marinetti,G.V., and O'Brien,D.F.(1978): <u>Biochemistry</u>, 17:
     4186-4192.

29. Degrip,W.J.,Daemen,F.J.M., and Bonting,S.L.(1973): Biochim. Biophys. Acta, 323:125-142.

30. Drenthe,E.H.S.,Bonting,S.L., and Daemen,F.J.M.(1980): Biochim. Biophys. Acta, 603:117-129.

31. Edel-Harth,S.,Dreyfus,H.,Bosch,P.,Rebel,G.,Urban,P.F., and Mandel,P. (1973): FEBS Lett., 35:284-288.

32. Frank,R.N.,Cavanagh,H.D., and Kenyon,K.R.(1973): J. Biol. Chem., 248:596-609.

33. Fung,B.K.K., and Hubbell,W.L.(1978): Biochemistry, 17:4396-4402.

34. Fung,B.K.K., and Hubbell,W.L.(1978): Biochemistry, 17:4403-4410.

35. Funk,J.,Welte,W.,Hodapp,N.,Wutschel,I., and Kreutz,W.(1981): Biochim. Biophys. Acta, 640:142-158.

36. Goldsmith,T.H.(1975): In: Photoreceptor Optics, edited by A.W. Snyder and R. Menzel, pp. 392-409. Springer-Verlag, New York.

37. Goldsmith,T.H., and Wehner,R.(1977): J. Gen. Physiol., 70:453-490.

38. Hagins,F.M.(1973): J. Biol. Chem., 248:3298-3304.

39. Hagins,W.A., and Liebman,P.A.(1963): Abstracts 7th Ann. Mt. Biophys. Soc.,ME 6. Statler-Hilton Hotel, New York City, N.Y. Feb. 18-20,1963.

40. Hall,M.O.,Bok,D., and Bacharach,A.D.E.(1969): J. Mol. Biol., 450:397-406.

41. Heller,J.(1968): Biochemistry, 7:2906-2913.

42. Heller,J.(1969): Biochemistry, 8:675-679.

43. Hess,H.H.,Stoffyn,P., and Sprinkle,K.(1976): J. Neurochem., 26: 621-623.

44. Heuser,J.E.(1977): In: Soc. for Neurosci. Symp., Vol. II, edited by W. M. Cowan, and J. A. Ferrendelli, pp. 215-239. Society for Neurosciences, Bethesda, Maryland.

45. Holm,M.,Mansson,J-E,Vanier,M-T., and Svennerholm,L.(1972): Biochim. Biophys. Acta, 280:356-364.

46. Kinney,M.S., and Fisher,S.K.(1978): Proc. R. Soc. Lond. (Biol), 201:149-167.

47. Krauhs,J.M.,Mahler,H.R.,Minkler,G., and Moore,W.J.(1976): J. Neurochem., 26:281-283.

48. Krauhs,J.M.,Mahler,H.R., and Moore,W.J.(1978): J. Neurochem., 30: 625-632.

49. Kuhn,H.(1980): Neurochemistry, 1:269-285.

50. Litman,B.J.(1974): Biochemistry, 13:2844-2848.

51. Lolly,R.N.(1980): In: Current Topics in Eye Research, Vol. 2, edited by J. A. Zadunasky and H. Davson, pp. 67-118, Academic Press, New York.

52. Mason,W.T., and Abrahamson,E.W.(1974): J. Membr. Biol., 15:383-392.

53. Mason,W.T.,Fager,R.S., and Abrahamson,E.W.(1973): Biochim. Biophys. Acta, 306:67-73.

54. Michel-Villaz,M.,Saibil,H.R., and Chabre,M.(1979): Proc. Natl. Acad. Sci. USA, 76:4405-4408.

55. Miljanich,G.P.,Sklar,L.A.,White,D.L., and Dratz,E.A.(1979): Biochim. Biophys. Acta, 552:294-306.

56. Molday,R.S.,and Molday,L.L.(1979): J. Biol. Chem., 254:4653-4660.

57. Nakano,T.,Ikai,A.,Nishigai,M., and Noda,H.(1979): J. Biochem., 85:1339-1346.

58. Nashima,K.,Mitsudo,M., and Kito,Y.(1979): Biochim. Biophys. Acta, 579:155-168.

59. Papermaster,D.S.,and Besharse,J.C.(1970): J.Cell Biol., 83:275a.

60. Papermaster,D.S.,Converse,C.A.,and Sin,J.(1975): Biochemistry, 14:1343-1352.

61. Papermaster,D.S.,Converse,C.A.,and Zorn,M.(1976): Exp.Eye Res., 23:105-115.

62. Papermaster,D.S., and Dryer,W.J.(1974): Biochemistry, 13:2438-2444.

63. Papermaster,D.S.,Schneider,B.G.,Zorn,M.A., and Kraehenbuhl,J.P. (1978): J. Cell Biol., 77:196-210.

64. Paulsen,R., and Schwermer,J.(1973): Eur. J. Biochem., 40:577-583.

65. Plantner,J.J., and Kean,E.L.(1976): Exp. Eye Res., 23:281-284.

66. Poo,M.M., and Cone,R.A.(1973): Exp. Eye Res., 17:503-510.

67. Poo,M.M., and Cone,R.A.(1974): Nature (London), 247:438-441.

68. Raubach,R.A.,Memes,P.O., and ·Dratz,E.A.(1974): Exp. Eye Res., 18:1-12.

69. Rohlich,P.(1976): Nature (London), 263:789-791.

70. Rothschild,K.J.,Sanches,R.,Hsias,T.L., and Clark,N.A.(1980): Biophys. J., 31:53-64.

71. Saari,J.C., and Bunt,A.N.(1980): Exp. Eye Res., 30:231-244.

72. Saibil,H.,Chabre,M., and Worcester,D.(1976): Nature (London), 262:266-270.

73. Schmidt,W.J.(1938): Kolloidzeitschrift, 85:137-148.

74. Schichi,H.(1971): J. Biol. Chem., 246:6178-6181.

75. Schichi,H.,Lewis,M.S.,Irreverre,F., and Stone,A.L.(1969): J. Biol. Chem., 244:529-536.

76. Smith,H.G.,Stubbs,G.W.,and Litman,B.(1975): Exp. Eye Res., 20:211-217.

77. Stone,W.L.,Farnsworth,C.C., and Dratz,E.A.(1979): Exp. Eye Res., 28:387-397.

78. Watts,A.,Volotovski,I.D., and March,D.(1979): Biochemistry, 18:5006-5013.

79. Weber,K.M. and Zinkler,D.(1973): In: Biochemistry and Physiology of Visual Pigments, edited by H. Langer, pp. 327-334. Springer-Verlag, New York.

80. Williams,D.S.,and Blest,A.D.(1980): Cell Tissue Res., 205:423-438.

81. Wolken,J.(1971): Invertebrate Photoreceptors, Academic Press, New York.

81a. Wolken,J.(1961): In: The Structure of the Eye, edited by G. K. Smelser, pp. 173-192. Academic Press, New York.

82. Wolken,J.(1974): In: The Eye: Comparative Physiology, edited by H. Davson and L. T. Graham, Jr. pp. 111-154. Academic Press, New York.

83. Wei,C-W., and Stryer,L.(1972): Proc. Natl. Acad. Sci. USA, 69:1104-1108.

84. Yamada,E.(1979): J. Electron Microsc., 28:S-79-S-86.

85. Yeager,M.,Schoenborn,B.,Engelman,D.,Moore,P. and Stryer,L.(1980): J. Mol. Biol., 137:315-348.

86. Young,R.W.(1974): Exp. Eye Res., 18:215-223.

87. Young,R.W.(1976): Invest. Ophthalmol., 15:700-725.

88. Zinkler,D.(1975): Verh. Dtsch. Zool. Ges., 67:28-32.

89. Zorn,M., and Futterman,S.(1971): J. Biol. Chem., 246:881-886.

Visual Cells in Evolution, edited by Jane A. Westfall,
Raven Press, New York © 1982.

Fine Structure and Turnover of Photoreceptor Membranes

Talbot H. Waterman

Department of Biology, Yale University, New Haven, Connecticut 06520

Eyes and extraoptic photoreceptors show a number .of striking
analogies with photographic and television cameras. Yet they differ
fundamentally in being open steady state systems far from Gibbsian ther-
modynamic equilibrium. This may not appear to affect their light gath-
ering and image forming properties critically but it clearly is basic to
the way their photoreceptive membranes are maintained in a state of continuous adaptive effectiveness.

As living things, and unlike photographic films or electronic sensors,
light responsive elements in animals are in a state of constant flux.
Information, energy and materials flow continually through the receptor
membrane. They respectively permit it to control excitation in the
visual system, to do physiological work and to maintain itself by
ongoing renewal.

The last of these processes is part of the central cellular problem
of membrane shuttling and recycling; it is the principal concern of
the present discussion. Vision, retinal maintenance and photoreceptor
membrane turnover are so intimately interdependent that they need to be
understood as a whole. Because this field is such a broad and currently
active one (e.g., 28, 44, 69, 74) a definitive review will not be
attempted here. Instead primary attention will be given to rhabdom-
bearing photoreceptors as well as their comparative relevance to turnover pathways and mechanisms being studied in many animals including
man and other vertebrates. Even this restricted field is sufficiently
lively and unsettled to warrant use of our own research on crustacean
photoreceptor membranes as a central focus for discussion (Figures 1-6).

THE STATUS QUO

Some of the basic principles of membrane turnover have been known for
a long time. Hence we have had occasion in the past to believe that our
understanding of such systems was reasonably good. But, as will become
clear below, a major generality emerging from current close interest in
these turnover problems (e.g., 6, 21, 32, 42, Waterman and Piekos, in
preparation) is that earlier optimistic notions were somewhat premature.
The same can be said for parallel research on vertebrate systems (e.g.,
1, 39, 50, 53, 71). Much additional experimental study is required,
ranging from membrane deployment of visual-pigment chromophores to
timing of turnover cycles by exogenous and endogenous factors. The
whole system depends, of course, on receptor membrane events. It also

extends to a whole hierarchy of interrelated elements: underlying cytoplasmic organelles for supplying and removing materials or components;
receptor cell nuclei for directing intracellular synthetic processes;
and intercellular communication and transfer mechanisms that in turn
are supported by organismal integrating elements such as the nervous and
endocrine systems.

Our interest in such processes has been strongly stimulated by three
striking phenomena found in decapod crustacean compound eyes. First is
the fact that nearly all cytoplasmic organelles in photoreceptor cells
are markedly affected by light and dark adaptation of the system (18).
Secondly, the area of photoreceptor membrane in a fully dark-adapted
eye can be nearly 20 fold greater than in the fully light-adapted state
(42). Thirdly, it has also become clear that the degree of adaptation
evoked by light or dark depends on the time of day (42) as it does in
vertebrates (e.g., 4, 38, 39, 43).

Arising as they did from an abiding interest in polarization sensitivity (55, 59), these three findings provide major themes for our research on crustacean compound eyes. Augmented by current experiments,
they also provide a topical basis for organizing this review.

EFFECTS OF LIGHT ON EYE FINE STRUCTURE

Research begun in the 1960's and continuing into the present has demonstrated that light and darkness affect the steady state of photoreceptor membranes in many vertebrates and invertebrates. Our work on fine
structural changes in the compound eyes of the brine shrimp Artemia,
the crayfish Procambarus and the spider crab Libinia was reviewed in a
1965 symposium (17), where both the disruptive effects of continuous
prolonged darkness (up to three months) and the cyclic effects of light
and dark adaptation were briefly reported.

Libinia experiments

Further experiments on Libinia eyes dark- or light-adapted for 5h and

FIG. 1. Distal third of isolated Carcinus retinula with tip partly dissociated to show four distal lobes of R_8 (Fig. 2) normally enclosed by
outer ends (Rn) of the regular retinular cells (R_1-R_7). Nomarski contrast. Rh, rhabdom. Original. Scale bar = 16 μm.

FIG. 2. Basal end of two Callinectes retinulas showing basement membrane (BM), hemolymph channel (HC) and basal attachment (BA) of rhabdom
(Rh) which is about 3 μm in diameter at its base. Nomarski contrast of
a 20 μm section. Waterman and Campbell, original.

FIG. 3. Isolated R_8 from a Carcinus retinula (Fig. 1). Four distal
lobes (L) bear the axial rhabdomere; one of them contains the cell's nucleus and continues proximally as an axon (A). In the figure this has
been broken near the basement membrane. Original. Scale bar=39 μm.

FIG. 4. Rhabdom (Rh) of Squilla showing its shape and relation to some
of the retinular cells (RC); perirhabdomal vacuole (PRV) crossed by many
protoplasmic bridges linking the photoreceptor membranes to their parent
cell cytoplasm. Scanning electronic micrograph: Waterman and Pooley,
original. Scale bar = 10 μm.

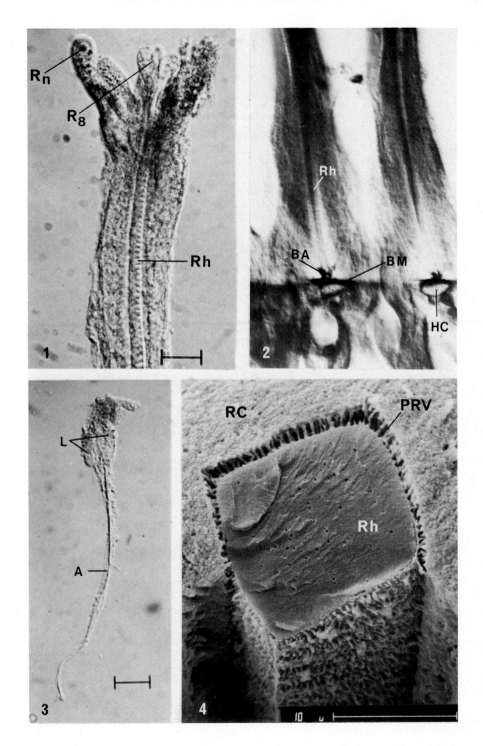

17h demonstrated profound cytological changes in the retinular cells
(18). The occurrence of numerous pinocytotic vesicles about 0.1 μm
in diameter and a series of larger lysosome-like vesicles was maximal
with the longer period of light adaptation and minimal with the longer
dark adaptation. It was suggested that these two kinds of cytoplasmic
organelles were part of an autophagic system that disposes of rhabdom
"metabolites" resulting from light stimulation (Fig. 7).

Intracellular vacuoles around the distal rhabdom and in the nearby
cytoplasm were also affected, but inversely, by light and dark. These
cisternae of smooth endoplasmic reticulum were maximal in size after
17h in the dark and minimal in volume and more dispersed after 17h in
the light. These responses are similar to those in the so-called palisade reported for Locusta (31), although the changes observed in Libinia
were less drastic. Perirhabdomal vacuoles in Artemia compound eyes were
not observed to enlarge on dark adaptation (26).

Both smooth and granular endoplasmic reticulum of the spider crab
were more extensively developed in light-adapted eyes than in those in
the dark. Conversion of vacuolar or tubular endoplasmic reticulum to a
lamellar form with flattened cisternae may account for the light induced
increase in the granular type. In addition, the number of free ribosomes in the cytoplasm increased significantly on light adaptation.
These observations were consistent with the idea that light activates
the photoreceptor cell membrane transport system as well as the protein
and lipid synthesizing cytoplasmic organelles.

Other observations in Libinia mainly confirmed typical light-related
migrations of proximal retinal pigment within the retinular cells and of
distal retinal pigment in pigmented glia around the distal retinula. A
new observation on proximal pigment movement was that light induced the
relatively uniform screening pigment sleeve around the light-adapted
rhabdom to divide into distally and proximally migrating fractions.

Golgi complexes, rarely observed in retinular cells, and mitochondria,
commonly seen proximally, were not found to change dramatically with
light or dark either in number or location.

Comparative data

Independently, and at about the same time, the photoreceptor cells of
larval mosquitoes (Aedes) were found (62, 63, 65) to show light related
cytoplasmic changes similar to those we found in adult compound eyes of
Libinia. It also should be noted that rod outer segment turnover in
vertebrates was being discovered about the same time (70, 73) as the
membrane cycling in Libinia and in Aedes larvae. One of the earliest reports in this general field was on photoreceptor membrane cycling in a

FIG. 5. Isolated rhabdom of Squilla demonstrating shape, alternately
orthogonal banding and distal location of rhabdomere of R8. Proximal
end down. Scanning electron micrograph: Waterman, Pooley and Piekos,
original. Scale bar = 10 μm.

FIG. 6. Retina of Carcinus shown in radial section nearly parallel to
retinular axes. Scanning electron micrograph of an etched 1 μm section.
BM, basement membrane; PRV, perirhabdomal vacuole; RC, retinular cell
cytoplasm; Rh, rhabdom. Waterman, Pooley and Piekos, original. Scale
bar = 10 μm.

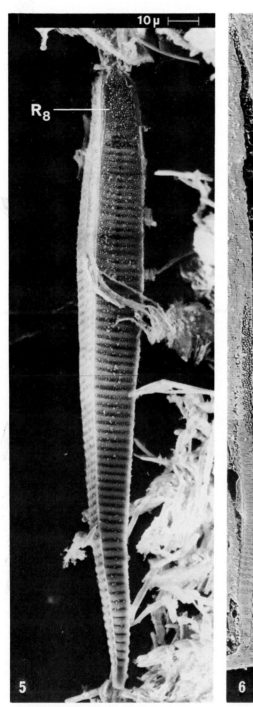

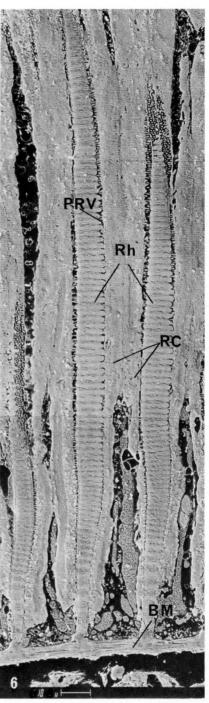

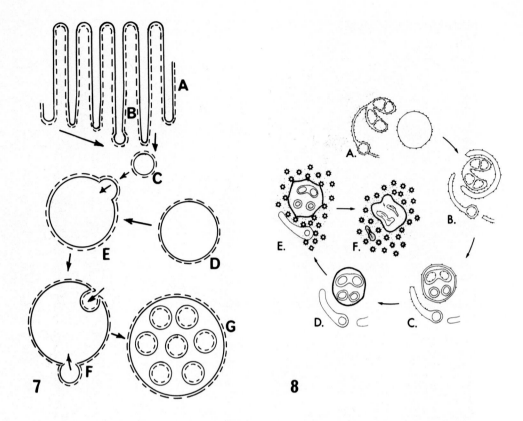

FIG. 7. Hypothesis for degradative turnover of rhabdom membrane in crayfishes. A: Part of a rhabdomere, showing five microvilli. B: Basal elongation between microvilli and the formation of a pinocytotic vesicle that gives rise to C, coated vesicle; D, large cytoplasmic vesicle; E, coated vesicle fusing with larger vesicle; and F, secondary endocytosis of large vesicle giving rise to a multivesicular body, G. (20).

FIG. 8. Hypothesis for membrane transformation in spider Dinopis retinular cells preparing for massive dawn photoreceptor membrane breakdown. Fenestrated rough endoplasmic reticulum (A) reorganizes (B) losing ribosomes (C) and synthesizing internal contents (D), producing numerous coated vesicles (E) to form a so-called Nebenkern (F). (10).

planarian flatworm (48).

As will be reviewed below, a number of additional arthropods, e.g., Artemia (26) and Leptograpsus (51), have since been studied in this regard so that comparative data are now available for a wide range of rhabdom-bearing eyes. However in the mid-sixties we were mainly interested in using these receptor cell cycles to identify retinal information channels particularly with relation to e-vector and wavelength discriminating mechanisms (19, 22, 41, 56, 58, 60).

Repeated study of such light-induced membrane turnover for other purposes increased our interest in the underlying cellular phenomena so that it has become a major research interest. The application of freeze-fracture electron microscopy to an analysis of retinular cell function

(20) provided better support for our earlier hypothesis of photoreceptor membrane degradation via pinocytotic vesicles and a sequence of lysosomes.

Freeze-fracture data on Procambarus demonstrated the presence of closely packed spheroidal particles on the protoplasmic face of fractured photoreceptor membranes in rhabdom microvilli. These elements, which apparently represent single (or small clusters of) rhodopsin molecules, also are present in the pinocytotic vesicles, multivescular bodies and lamellar bodies in the degradative sequence. In later stage lysosomes the density of such membrane-borne particles is significantly reduced.

In retinular cell membrane remote from the rhabdom the particles are absent or, if present, they are less concentrated and less uniform in size. Tests for the presence of acid phosphatases were negative in early stage multivesicular bodies, but positive in later stages that could thereby be identified as secondary lysosomes. More recent work on the crab Leptograpsus has shown that such hydrolases originate both from structures identified as Golgi bodies and from rough endoplasmic reticulum (12).

Membrane lysis apparently similar to that in decapod crustaceans has also been reported in spiders where very rapid dawn breakdown coupled with comparably rapid dusk synthesis occur in Dinopis' posterior median eyes (5, 9, 11). Cytological evidence has been presented for an elaborate predawn hydrolase synthesis system (Fig. 8).

Extracellular degradation

Whereas endocytosis at the base of rhabdom microvilli followed by lysosomal degradation is widely present in various crustaceans, spiders (e.g., Menneus, 13) and insects, other pathways of membrane destruction also occur. A somewhat different alternative route was reported in the shrimp Palaemonetes (33) where localized rhabdom areas appeared to form membrane whorls that moved into the neighboring cytoplasm as lamellar bodies. These in turn were presumed to transform into dense bodies through lysosomal action. This pathway was alternative to the previously described pinocytotic vesicle-multivesicular body-lamellar body route.

More divergent extracellular pathways have also been found. In salticid spiders pinocytotic vesicles have not been found in the anterior lateral eyes of Plexippus; multivescular bodies are absent in this form too (8) and are rare in other salticids (16). Rhabdomere membrane is endocytosed relatively slowly in Plexippus by adjacent glial cells, in which lysis occurs. Evidence for relations between receptor cells and retinal glial cells similar to those in Plexippus had been obtained earlier in the pisaurid spider Dolomedes (7).

The recycling mechanisms of these particular spiders are therefore not intracellular, as in the examples discussed above, but extracellular. In this regard it is reminiscent of the phagocytosis of shed outer segment fragments by fixed macrophages of the retinal pigment epithelium that occurs in many vertebrate retinas (e.g., 15, 35, 72). Rapid extracellular shedding has also been demonstrated in the crane fly Ptilogyna (68). After dawn much of the photoreceptor membrane breaks off into the extracellular space surrounding the rhabdomeres of this dipteran's open rhabdom.

The resulting extracellular membrane detritus is then engulfed by pseudopodia of the retinular cells themselves. Large multivescular

bodies form intracellularly from this material and apparently are lysed
through a subsequent lamellar and dense body sequence. In another fly,
Lucilia, coated pits in the receptor cell plasma membrane take up the
debris shed extracellularly at rhabdom breakdown (6). We have recently
found evidence in crustaceans for the participation of fixed glial
macrophages and of wandering hemocytes in still another case of extracellular degradation of arthropod photoreceptor membrane elements.

Retinal hemocytes

In the blue crab Callinectes work in progress indicates that there
are large quantities of detritus in the proximal cytoplasm of pigment
cells surrounding the retinula near the basement membrane (Toh and
Waterman, in preparation). These intracellular inclusions are phasic
in occurrence, and their appearance closely follows the period of intense
rhabdom breakdown occurring near dusk. At the same time large numbers
of motile macrophages are present in close proximity to retinular cells
with disintegrating rhabdomeres. Again the occurrence of detritus within
the hemocytes' cytoplasm is in phase with the receptor membrane disinte-
gration and the disappearance of the resulting retinal debris.

Whereas the rhabdom breakdown seems to be a much less catastrophic
process in the superposition eye of Procambarus, sustained presence of
hemocytes in their retinas is also correlated with membrane breakdown
(Waterman and Piekos, in preparation). Apparent phagocytosis of extracellular debris by these macrophages, like that reported in the breakdown
of spider (7, 8) and dipteran insect (6, 68) membrane systems, has been
observed, as has the formation of gap junction-like contacts with photoreceptor cell surface membrane. Observation of extracellular debris
within the cytoplasm of fixed glial cells in the Procambarus retina
suggests still another route of receptor membrane disposal.

Overall counts on retinal hemocytes show that in the dark a significant flow of hemocytes from extraretinal blood vessels into the retinular level of the retina occurred. Conversely, in the light these
macrophages emigrated below the basement membrane and hence out of the
retina. Experiments with one eye exposed to light and the other covered
proved that such emigration of hemocytes could be unilaterally induced
by light. Highest densities of hemocytes were counted just proximal to
the basememt membrane, but their greatest light-induced change in density
occurred in the retina just distal to that boundary layer.

Degradative pathways

Earlier work suggested (6) that some arthropods (e.g., Grapsus,
Libinia, Leptograpsus, Dinopis, Aedes larvae and adults) depend mainly
on an autophagic lysosomal degradation system for photoreceptor membrane
cycling. Others like Plexippus, Ptilogyna and Lucilia appear to depend
primarily on mechanisms outside the receptor cell to dispose of discarded
membrane. Our current observations on Callinectes demonstrate well-
developed autophagic as well as glial and hemocytic macrophagic lysing
pathways. However, their possible division of labor and relative importance cannot yet be evaluated.

More quantitative data are needed in nearly every instance before an
adequate assignment of importance can be made to the various mechanisms
at work. This is particularly true of crayfishes, even though their
retinas and visual physiology have been widely studied (e.g., 22, 25,

Cummins and Goldsmith, in preparation; Waterman and Piekos, in preparation).

Membrane synthesis

Whereas the preceding discussion purported to analyze photoreceptor membrane turnover in general, most of the data reviewed so far relates to membrane breakdown and removal. This phase of the cycle seems to be triggered or induced by light even though other factors are certainly involved, as will be discussed below. Little has been said thus far on the synthetic aspects of the turnover required to counteract ongoing breakdown or to rebuild membrane components removed by rapid degradation. Actually there is less known about the production and assembly of receptor membranes than their disassembly.

This is probably true for vertebrate outer segments, in spite of the classic labelling of rod disk serial replacement by Young (70, 71), and it is certainly so for rhabdomal systems where membrane-labelling techniques have generally proved less productive. Rhabdom growth and assembly is clearly correlated with the onset of darkness, but the absence of light is only part of the control mechanism. In several instances the possibility that the numerous coated vesicles and multivesicular bodies observed might be of two types, one of them involved in synthesis, either has not been demonstrated or has been shown not to be true (5, 9). In our studies on many crustacean rhabdoms the finding most relevant to synthesis was in a freshwater ostracod eye (20) and in the compound eyes of the isopods Asellus and Cleantiella (Eguchi and Waterman, in preparation). There, extensive connections were demonstrated between microvillar membranes and smooth endoplasmic reticulum tubules in the adjacent cytoplasm. In other areas of retinular cells numerous pinocytotic vesicles and multivescular bodies typical of the membrane-degradative system were found. The clear distinction between the relationship of microvilli to smooth endoplasmic reticulum and to autophagic lysing organelles suggests the involvement of the former with photoreceptor membrane synthesis. Indeed the importance of smooth endoplasmic reticulum in photoreceptor membrane synthesis was suggested in the same year by Itaya (33) and Whittle (66) and subsequently supported by Blest and Day (7) and Stowe (51, 52).

Itaya (33) hypothesized a model for the complete cycle of rhabdom membrane synthesis and breakdown in the shrimp Palaemonetes. This included a special type of degradation cited above as well as two possible synthetic routes for which some transmission electron microscopical evidence was presented. For one of these routes double membranes of smooth endoplasmic reticulum were shown to be continuous with proximal photoreceptor membrane at the base of rhabdomere microvilli, apparently paralleling our ostracod and current isopod data. In the other pathway proposed, smooth endoplasmic reticulum condensed into whorls of double membranes called concentric ellipsoids formed the critical organelles involved. These, along with simple membrane vesicles, were closely aggregated adjacent to regions of intact rhabdomere. The membrane aggregates were believed to be in the process of organizing into the highly ordered microvillus pattern of a fully differentiated rhabdomere.

As we later documented quantitatively for Grapsus (42) the rhabdom's diameter increase in the dark adapting Palaemonetes retina depends on a doubling of microvillar lengths in one transverse axis and a substantial increase in the number of microvilli present along the orthogonal

transverse axis (33). The two paths proposed for receptor organelle
synthesis were modeled to account respectively for the two directions
of rhabdom growth.

In her ongoing work on Leptograpsus membrane cycling, Stowe (51,
52) has developed a concentric ellipsoid model as the major mechanism
of microvillus regeneration in the retina of this grapsoid rock crab.
Before discussing membrane synthesis further it seems desirable to in-
troduce two other dimensions of photoreceptor membrane turnover that are
essential for understanding the present status of the problem. They are
the balance between synthesis and degradation and its changes with time.
Obviously these factors determine the membrane area at any given moment.
Such dynamic characteristics of the receptor system obviously involve
one of its most interesting aspects, namely its control.

MEMBRANE AREA CHANGES

As open systems, photoreceptor membranes are able to maintain their
stability as long as their input and output are appropriately regulated.
In other words, if synthesis exactly matches breakdown the size and form
of the light sensitive organelles will remain the same even though their
substance is constantly changing at rates that may vary over a wide
range.

If input exceeds output the membrane area will increase and vice
versa; if local differences in input-output occur it will change shape.
Both kinds of imbalance may appear in rhabdoms and their analysis provides an important means of understanding membrane turnover and its
control. In the absence of size and shape changes other quantitative
techniques are required (e.g., 14, 24, 27, 47, 70).

Where quantitative data are available evidence indicates that turnover rates are extraordinarily high in certain arthropods (e.g., 64).
Particularly striking results have been reported for membrane area
changes in Dinopis (5) and Grapsus (42). The case of Grapsus, in which
there is a difference of nearly 20 fold in photoreceptor membrane area
between midnight dark-adapted and noon light-adapted eyes, may be considered an extreme example (Fig. 9).

As mentioned above, the underlying expansion and contraction of the
rhabdom involves mainly elongation and shortening of its constituent
microvilli along their axes and substantial changes in their number at
90° to the axial direction; relatively minor differences occur in microvillus diameter as well as in the thickness and number of bands making
up the length of the rhabdom.

Whereas appropriate experiments have yet to be done, the functional
effect of a 16 fold change in the rhabdom's cross section can be estimated from the optical geometry (29, 30). Other things being equal,
sensitivity to an extended source would be increased in the midnight
dark-adapted rhabdom by a factor corresponding to the growth in cross
section but there would be some associated loss in the resolution of
visual detail. The amplitude of change would depend on the relations
between interommatidial angles, the diameter of the Airy disk and the
stop diameter (here assumed to be determined by the diameter of the
rhabdom tip) in the focal plane of the ommatidium in question.

Actually we know that visual adaptation even at the retinal level is
a complex process (e.g., 57). Consequently our emphasis here on the
fine structure and area of the photoreceptor membrane neglects for the
moment several important factors that interact with it. One of the

most basic and interesting of these is density and specific deployment
of visual pigment molecules and their chromophores that may change
quite independently of the membrane area.

For example, in crayfish genetically deficient in normal screening
pigment, light-exposure caused not only a decrease in rhabdom area but
also a 10-20 fold decrease in the areal density of rhodopsin molecules
and a resultant 70 fold loss of sensitivity (34). Visual discrimination
in general will certainly be strongly dependent on the interaction between these and other relevant factors.

In the wide variety of eyes that have been studied changes of photoreceptor membrane diameter correlated with light and dark adaptation
range from essentially zero in the vertebrates to 4 fold in Grapsus
(Fig. 9). Actually photomechanical movements of receptor components are
often present (Figs. 10-12), and pigment migration or other entrance
pupil determining factors usually play an important secondary role in
controlling visual function (40). However, our main attention here is
on the photoreceptor membrane itself.

Within the arthropods and even within their major classes, such as
Crustacea and Insecta, there is a lot of variation. Thus in response
to light and dark adaptation Grapsus rhabdom volume changes by a factor
of nearly 20 fold, Dinopis by about 10 fold (5), Artemia around 6 fold
(26), Leptograpsus by 2-3 fold (51, 52), Callinectes by 2-3 fold (Toh
and Waterman, in preparation) and Procambarus by only 25% in maximum
diameter (25' Waterman and Piekos, in preparation).

Blest (6) has suggested that these diurnal changes in photoreceptor
organelle size are correlated with two types of compound eyes. Those
like Dinopis, Grapsus, Aedes, and Limulus (3) seem to correlate well
with the presence of retinular cells having numerous pinocytotic vesicles and well-developed lysosomal systems. Such receptors would appear
to be adapted for nocturnal vision with high absolute sensitivities or
for a wide range of environmental light intensities (e.g., Grapsus). The
second type proposed is typified by lack of marked diurnal changes in
rhabdom size and by retinular cells that have few or no pinocytotic
vesicles and lysosomes in their retinular cytoplasm. In this class are
the lateral ocelli of Plexippus mentioned above and the compound eyes
of the diperans Lucilia and Calliphora. These are primarily diurnal in
function and have a low absolute sensitivity.

MEMBRANE CYCLES IN TIME

Earthbound organisms live predominantly on a 24h schedule dominated
by a corresponding program of sunlight and darkness. This has a profound
effect on the whole organism that is especially dramatic with regard
to vision. Photoreceptor membranes are formed and destroyed at rates
directly related to changes in light intensity that trigger or modulate
their breakdown and synthesis. In addition, at least certain aspects
of their activity are endogenously controlled by some components of the
biological clock.

The earliest persistent endogenous rhythm in crustacean eyes reported
by Welsh (61) was variation in the levels of the distal retinal screening
pigment in the shrimp Macrobrachium. Subsequent work showed that such
rhythms in pigment migration occurred in various arthropods and could
persist for several months under apparently constant conditions. In
many species rhythms are maintained in continuous darkness, continuous
light or both. One or more of the several screening pigments present

may show such rhythmic behavior in various instances. In addition, many studies have shown that the quantitative effect of a given light- or dark-adaptation period varies in a systematic manner with time of day.

An introduction to the complex neuroendocrine origins of such rhythms may be found in recent crayfish research (2, 37). Isolated eyestalks also may continue to show persistent rhythms of proximal pigment movements and of ERG amplitude (45, 49). The effect of various periods of adaptation on rhabdom size in Grapsus showed that the effects of the same light regime were quantitatively different at various times of day (42).

Similarly, current experiments on Callinectes (Toh and Waterman, in preparation) indicate that initiation of strong synthetic or degradative changes in the rhabdom depend on marked light intensity shifts, at dusk and dawn respectively. On the other hand the readiness of retinular cells to carry out these acute phases of membrane turnover does not undergo the changes expected if there is a circadian rhythm involved. This has been demonstrated in crabs that have been maintained for several days in constant darkness (D:D) after extended entrainment to a laboratory imposed regime of L:D in phase with local day:night sequence. In such animals the normal transition of the rhabdom's nocturnal maximal size to its minimal midday state does not occur. Hence the accelerated shedding associated with dawn and morning is absent. Instead the photoreceptor membrane continues through the dark day period to show a

———————————▶

FIG. 9. Comparison of noon light-adapted (LA) and midnight dark-adapted (DA) retinulas of the crab Grapsus. The rhabdom (Rhab) length changes from 289 μm to 336 μm while its diameter grows from 1.99 μm to 7.89 μm. Hence photoreceptor membrane surface area increases by nearly 20 fold. ARC, axons of regular retinular cells; AR_8, axon of R_8; BM, basement membrane; CS, proximal end of crystalline cone stalk; NRC, nucleus of regular retinular cell; NR_8, nucleus of R_8; R_1-R_7, regular retinular cells, one through seven; R_8, retinular cell eight. (42).

FIG. 10. Comparison of day-adapted (left) and night-adapted (right) receptor cells from posterior median eyes of the spider Dinopis. Each cell is hexagonal in cross section with a rhabdomere unit on each face. Dark adaptation markedly lengthens rhabdomere microvilli at the expense of receptor cell cytpolasm and the receptor segment elongates by about 3 fold. The overall growth in photoreceptor membrane is about 10 fold. Note that photomechanical movement of screening pigment (black rectangles) is another prominent change. (5).

FIG. 11. Comparison of night-adapted (A) and day-adapted (B) retinulas in the tipulid fly Ptilogyna. Modest increases in rhabdomere diameters and length result in an increase in rhabdom volume of 2-3 fold in the period between noon and midnight. Photomechanical movements are extensive not only for the retinular cells and rhabdom but also for primary and secondary pigment cells as well as the cone tract. (67).

FIG. 12. Comparison of light-adapted (left) and dark-adapted (right) cone cells from the retina of the teleost Haemulon. Photomechanical myoid elongation and decrease in microtubule numbers are the main changes; B, basal body of cilium; F, microfilaments; M, mitochondria of ellipsoid; OLM, outer limiting membrane; MT, microtubules; N, nucleus. (54).

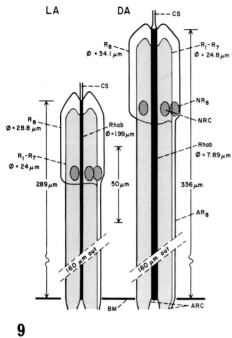

9

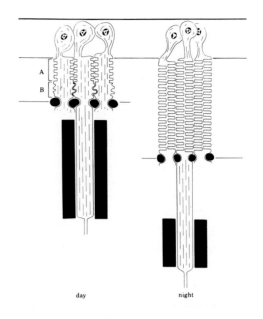

10

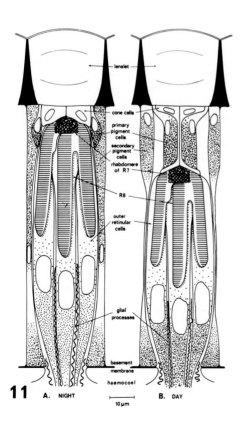

11

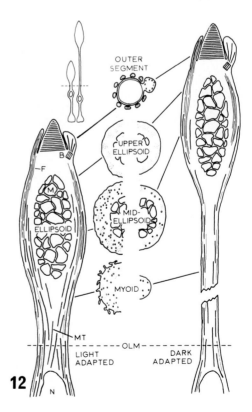

12

substantial increase in size beyond the maximum normally achieved. This accelerated synthesis is already evident in the dark one hour after dawn would ordinarily have occurred.

Peak rhabdom size was achieved at midday on the first day in the dark and was maintained on the second day in continuous darkness. Thus in the absence of a light-dark transition at dusk accelerated synthesis occurred rather than accelerated breakdown leading to the normal mid-night rhabdom size. However some cytoplasmic elements apparently continued in D:D to follow their normal pattern at least in part. For example, the smooth endoplasmic reticulum shows some circadian changes under D:D; multivescular bodies also are "spontaneously" rhythmic because they become more numerous by the end of the first hour after normal dawn in the D:D sequence. Also the perirhabdomal vacuoles became smaller on a similar D:D schedule thus following to that extent their normal L:D behavior.

In Leptograpsus (52) as well as in some spiders (5, 6), alterations in the normal L:D cycle induces substantial changes in the photoreceptor membrane turnover schedule. Early darkness correspondingly advanced the synthetic phase normally present in early evening hours for Leptograpsus and the resulting rhabdom was subnormal in size. Painting over both eyes during the light period of the L:D cycle had similar effects to those resulting from advancing the dusk. When this was done unilaterally the rhabdom of the painted eye became significantly larger than that of the unpainted one. The dark induced synthesis had no apparent effect on the contralateral eye. Continuing light past the normal time of dusk resulted in rhabdom diameters smaller than those normal for early evening.

With the data currently available it is difficult to describe a mechanism through which light and darkness act to trigger the observed membrane effects. Two facts evident in the Leptograpsus and Callinectes data epitomize this problem. At a given moment receptor cells from one retina or even one retinula may be in quite different phases of the overall cycle. The other is that given cellular phases appear in various retinular cells at different times in the L:D cycle spread over a 2-3 hour period. In the early post dusk period when, on an average, peak synthetic activity occurs in these crabs, some cells appear to have already completed full rhabdom growth while others seem barely to have begun. Obviously an effective model must account for such evidence of loose coupling between the apparent external signal and the membrane cycling response. Better quantitative data is certainly needed to make effective progress here.

In our present incomplete state of understanding there may be an emerging difference of opinion about the continuous or discontinuous nature of photoreceptor membrane turnover in rhabdoms. This may, however, mainly depend on real differences between the various species studied. In Palaemonetes for example the breakdown and synthesis of membrane were reported to be two discontinuous processes triggered by dawn and dusk changes in light intensity (33). No detailed or quantitative data were presented to support this conclusion.

A somewhat parallel turnover schedule was well documented for Dinopis (5) in which the size changes occurred mainly within the first two hours following light changes at dawn and dusk. Normal breakdown did not occur in the morning hours if darkness was continued, nor was normal rapid synthesis present if illumination was continued through the evening. Both an endogenous rhythm and direct effects of light and darkness were postulated to be responsible for the observed turnover schedule.

The most extreme case of a catastrophic turnover system has been reported for Leptograpsus (51, 52). In this animal it is argued that there is an extensive dissolution of the photoreceptor membranes that normally occurs soon after dusk every day. Whole rhabdomeres are postulated to breakdown at this time to be replaced by freshly synthesized membrane arriving via the tubular endoplasmic reticulum-concentric ellipsoid system cited above. Whereas quantitative data have not yet been presented on this Leptograpsus turnover, the proposed mechanism involves the addition of a drastic new step in the program of continuous changes in rhabdom size correlated adaptively with ambient light intensities and their changes. This new step is a large scale degradation of old membrane before or in the course of the continuous growth of rhabdom size that marks the early evening.

In contrast to such episodic membrane changes our evidence for Grapsus (42) led us to suggest a sinusoidal function for the relation between rhabdom size and time of day. Obviously the imbalance in turnover rates for such a function would be maximal near the dawn and dusk periods but would be minimal or approach zero near the noon and midnight minimal and maximal sizes.

Our current work on Callinectes (Toh and Waterman, in preparation demonstrates that its rhabdom reaches maximum diameter before midnight and thereafter diminishes somewhat to an intermediate plateau. Our transmission electron micrographs of eyes fixed in the early evening show some rhabdomeres in a thoroughly disorganized state resembling those figured by Stowe (52) for Leptograpsus. We do not have any measure yet of how extensive these apparently dissociating rhabdomeres may be in the retina or even how much of the total rhabdomere is involved in the dissociated area. In our parallel observations on Procambarus (Waterman and Piekos, in preparation) no such extensive regions of photoreceptor disorganizations have yet been observed. However, the evidence that such massive breakdowns occur in a narrow time slot (perhaps within 30-60 min) means that special care needs to be taken that data are collected at the right time of day and under appropriate conditions of natural or adequate simulated illumination (32). We are continuing to extend our studies on both Callinectes and Procambarus in ways that should resolve some of these uncertainities.

Although a comparative and hopefully eclectic approach has been used above, no explicit reference has been made to the central topic of this book, photoreceptor cell evolution (see especially chapters by Eakin, Vanfleteren and Salvini-Plawen, this volume). The lack of a sufficiently detailed and coherent body of relevant data on the topic here reviewed inhibits the derivation of a broadly based construct. This is however one of the long range objectives of our work. No doubt eyes and their constituent cells will provide an important source of evolutionary information. Some stimulating essays on this topic have already been made within the higher Crustacea (23, 36) and within the Arthropoda (46). Further progress should be forthcoming, particularly when more information is available on adaptive convergence in membrane turnover.

ACKNOWLEDGMENTS

Support for the author's current research is being provided by NIH Grant EY-02929. Thanks are due to Jean Kashgarian and Barry Piekos for their help in preparing this manuscript.

REFERENCES

1. Anderson,D.H.,Fisher,S.K.,Erickson,P.A., and Tabor,G.A.(1980); Exp. Eye Res., 30:559-574.

2. Aréchiga,H.(1977): Fed. Proc., 36:2036-2041.

3. Barlow,R.B.,Chamberlain,S.C., and Levinson,J.Z.(1980): Science, 210:1037-1039.

4. Basinger,S.,Hoffman,R., and Mathes,M.(1976): Science, 194:1074-1076.

5. Blest,A.D.(1978): Proc. R. Soc. Lond. (Biol.), 200:463-483.

6. Blest,A.D.(1980): In: The Effects of Constant Light on Visual Processes, edited by T.P. Williams and B.N. Baker, pp. 217-245. Plenum, New York and London.

7. Blest,A.D., and Day,W.A.(1978): Philos. Trans. R. Soc. Lond. (Biol.), 283:1-23.

8. Blest,A.D., and Maples,J.(1979): Proc. R. Soc. Lond. (Biol.), 204: 105-112.

9. Blest,A.D.,Kao,L., and Powell,K.(1978): Cell Tissue Res., 195:425-444.

10. Blest,A.D.,Powell,K., and Kao,L.(1978): Cell Tissue Res., 195:277-297.

11. Blest,A.D.,Price,G.D., and Maples,J.(1979): Cell Tissue Res., 199: 455-472.

12. Blest,A.D.,Stowe,S., and Price,G.D.(1980): Cell Tissue Res., 205: 229-244.

13. Blest,A.D.,Williams,D.S., and Kao,L.(1980): Cell Tissue Res., 211: 391-403.

14. Brammer,J.D.,Stein,P.J., and Anderson,R.A.(1978): J. Exp. Zool., 206:151-156.

15. Dowling,J.E., and Gibbons,L.R.(1962): J. Cell. Biol., 14:459-474.

16. Eakin,R.M., and Brandenburger,J.L.(1971): J. Ultrastruct. Res., 37:616-663.

17. Eguchi,E., and Waterman,T.H.(1966): In: The Functional Organization of the Compound Eye, edited by C.G. Bernhard, pp. 105-124. Pergamon, Oxford.

18. Eguchi,E., and Waterman,T.H.(1967): Z. Zellforsch. Mikrosk. Anat., 79:209-229.

19. Eguchi,E., and Waterman,T.H.(1968): Z. Zellforsch. Mikrosk. Anat., 84:87-101.

20. Eguchi,E., and Waterman,T.H.(1976): Cell Tissue Res., 169:419-434.

21. Eguchi,E., and Waterman,T.H.(1979): J. Comp. Physiol., 131:191-203.

22. Eguchi,E., Waterman,T.H., and Akiyama,J.(1973): J. Gen. Physiol., 62:355-374.

23. Fincham,A.A.(1980): Nature (London), 287:729-731.

24. Hafner,G.S., and Bok,D.(1977): J. Comp. Neurol., 174:397-416.

25. Hafner,G.S.,Hammond-Soltis,G., and Tokarski,T.(1980): Cell Tissue Res., 206:319-332.

26. Hertel,H.(1980): Zool. Jahrb. Abt. Allg. Zool. Physiol. Tiere, 84:1-14.

27. Hollyfield,J.G., and Rayborn,M.E.(1979): Invest. Ophthalmol., 18:117-132.

28. Holtzman,E., and Mercurio,A.M.(1980): Int. Rev. Cytol., 67:1-67.

29. Horridge,G.A.(1978): Phil. Trans. R. Soc. Lond. (Biol.), 285:1-59.

30. Horridge,G.A.(1980): Proc. R. Soc. Lond. (Biol.), 207:287-309.

31. Horridge,G.A., and Barnard,P.B.T. (1965): Q.J. Microsc. Sci., 106:131-135.

32. Horridge,G.A., and Blest,A.D.(1980): In: Insect Biology in the Future, edited by M. Locke and D.S. Smith, pp. 705-733. Academic Press, New York.

33. Itaya,S.K.(1976): Cell Tissue Res., 166:265-273.

34. Kong,K.L., and Goldsmith,T.H.(1977): J. Comp. Physiol., 122:273-288.

35. Kuwabara,T.(1979): In: The Retinal Pigment Epithelium, edited by K.M. Zinn and M.F. Marmor, pp. 58-82. Harvard University Press, Cambridge, Massachusetts and London.

36. Land,M.F.,Burton,F.A., and Meyer-Rochow,V.B.(1979): J. Comp. Physiol., 130:49-62.

37. Larimer,J.L., and Smith,J.T.F.(1980): J. Comp. Physiol., 136:313-326.

38. LaVail,M.M.(1976): Science, 194:1071-1074.

39. LaVail,M.M.(1980): Invest. Ophthalmol., 19:407-411.

40. Miller,W.H.(1979): In: Handbook of Sensory Physiology, Vol. VII/6A, edited by H. Autrum, pp. 69-143. Springer-Verlag, Berlin, Heidelberg, New York.

41. Nassel,D.R., and Waterman,T.H.(1977): <u>Brain Res.</u>, 130:556-563.

42. Nassel,D.R., and Waterman,T.H.(1979): <u>J. Comp. Physiol.</u>, 131: 205-216.

43. O'Day,W.T., and Young,R.W.(1978): <u>J. Cell. Biol.</u>, 76:593-604.

44. Olive,J.(1980): <u>Int. Rev. Cytol.</u>, 64:107-169.

45. Olivo,R.F., and Larsen,M.E.(1978): <u>J. Comp. Physiol.</u>, 125:91-96.

46. Paulus,H.F.(1979): In: <u>Arthropod Phylogeny</u>, edited by A.P. Gupta, pp. 299-383. Van Nostrand Reinhold, New York.

47. Perrelet,A.(1972): <u>J. Cell Biol.</u>, 55:595-605.

48. Röhlich,P., and Török,L.J.(1962): <u>O.J. Microsc. Sci.</u>, 104:543-548.

49. Sanchez,J.A., and Fuentes-Pardo,B.(1977): <u>Comp. Biochem. Physiol.</u> 56:601-605.

50. Steinberg,R.H.,Fisher,S.K., and Anderson,D.H.(1980): <u>J. Comp. Neurol.</u>, 190:501-518.

51. Stowe,S.(1980): <u>Cell Tissue Res.</u>, 211:419-440.

52. Stowe,S.(1981): <u>J. Comp. Physiol.</u>, 142:19-25.

53. Tabor,G.A.,Fisher,S.K., and Anderson,D.H.(1980): <u>Exp. Eye Res.</u>, 30:545-557.

54. Warren,R.H., and Burnside,B.(1978): <u>J. Cell. Biol.</u>, 78:247-259.

55. Waterman,T.H.(1950): <u>Science</u>, 111:252-254.

56. Waterman,T.H.(1966): In: <u>Proceedings of the Symposium on Information Processing in Sight Sensory Systems</u>, edited by P.W. Nye, pp. 48-56. California Institute of Technology, Pasadena.

57. Waterman,T.H.(1975): <u>J. Exp. Zool.</u>, 194:309-343.

58. Waterman,T.H.(1977): In: <u>Identified Neurons and Behavior of Arthropods</u>, edited by G. Hoyle, pp. 371-386. Plenum, New York.

59. Waterman,T.H.(1981): In: <u>Handbook of Sensory Physiology</u>, Vol. VII/6B, edited by H. Autrum, pp. 281-469. Springer-Verlag, Berlin, Heidelberg, New York.

60. Waterman,T.H., and Horch,K.W.(1966): <u>Science</u>, 154:467-475.

61. Welsh,J.H.(1930): <u>Proc. Natl. Acad. Sci. USA</u>, 16:386-395.

62. White,R.H.(1967): <u>J. Exp. Zool.</u>, 166:405-426.

63. White,R.H.(1968): <u>J. Exp. Zool.</u>, 169:261-278.

64. White,R.H., and Lord,E.(1975): J. Gen. Physiol., 65:583-598.

65. White,R.H., and Sundeen,C.D.(1967): J. Exp. Zool., 164:461-478.

66. Whittle,A.C.(1976): Zool. Scripta, 5:191-206.

67. Williams,D.S.(1980): Zoomorphologie, 95:85-104.

68. Williams,D.S., and Blest,A.D.(1980): Cell Tissue Res., 205: 423-438.

69. Williams,T.P., and Baker,B.N., editors (1980): The Effects of Constant Light on Visual Processes, 455 pp., Plenum, New York and London.

70. Young,R.W.(1967): J. Cell. Biol., 33:61-72.

71. Young,R.W.(1978): Vision Res., 18:573-578.

72. Young,R.W.(1979): In: Ophthalmology, (XXIII Internatl. Congr.), edited by K. Shimizu, pp. 159-166. Excerpta Medica, Amsterdam.

73. Young,R.W., and Droz, B. (1968): J. Cell. Biol., 39:169-184.

74. Zinn,K.M., and Marmor,M.F., editors (1979): The Retinal Pigment Epithelium, Harvard University Press, Cambridge, Massachusetts and London.

Visual Cells in Evolution, edited by Jane A. Westfall,
Raven Press, New York © 1982.

Morphogenesis of Photoreceptor Cells

O. Trujillo-Cenóz

*Laboratory of Comparative Neuroanatomy, Instituto de Investigaciones Biológicas, Clemente Estable,
Montevideo, Uruguay*

When examining the development of photoreceptors or when exploring
the phylogenetic relationships of diverse types of eyes, only those cell
structures directly involved in phototransduction have in most cases,
commanded attention. Nevertheless, it is well established that each
photoreceptor cell has, in addition to its transducer pole, a transmitter
pole synaptically linked to a more or less complex chain of neurons. It
is also generally admitted that the most sophisticated functional properties of an eye are subserved by these neural circuits. Therefore, in
this kind of study it is advisable to analyze in detail, not only the
ontogenic maturation of the transducer structures but also to devote
special attention to the development of the synaptic pole and its associated nervous components.

In order to achieve these goals, it is of paramount importance to
find some sort of ideal organism possessing a relatively simple eye suitable for both anatomical and experimental investigations. Considered in
this context, the compound eye of muscoid dipterans has proven to be a
fruitful biological material. Despite the fact that each compound eye
contains several thousands of photoreceptors, the exploration of the
maturation process is greatly aided by the modular organization of the
future sensory cells. These peripheral modules (cell clusters or protoommatidia) are identifiable at a relatively early developmental stage.
Moreover, the modular organization is also present at the level of the
visual neuropiles in such a way that the main synaptic units are easily
recognizable even under the light microscope. This kind of eye offers
two more experimental advantages: (1) the eye primordia can be used for
transplantation experiments, and (2) there is in Drosophila an increasing number of eye-mutants allowing the genetic dissection of some of
the biochemical and neural mechanisms involved in vision.

The present paper concerns the morphogenesis of photoreceptor cells,
giving special emphasis to the information derived from our investigations on the compound eye of the green bottle-fly Phaenicia sericata.
To facilitate understanding of developmental changes occurring in the
eye primordium, I shall provide some basic information on the structural
organization of mature, adult eyes. Hence I have included here, a brief
anatomical review covering the most typical histological features
characterizing the dipteran eye.

THE ADULT EYE. ITS MAIN STRUCTURAL CHARACTERISTICS

The dipteran eye is characterized by the following distinctive features (Fig. 1):

1. The ommatidia are of the open-type and therefore the rhabdomeres behave as morphologically and optically independent structures (5,8,9, 11,12).

2. A cross section passing through a single ommatidium shows six rhabdomeres distributed according to a trapezoidal pattern (9). There is also a seventh centrally situated rhabdom, consisting of two coaxial rhabdomeres derived from the superior and the inferior central cells (cells 7-8) (14).

3. The eight axons arising from the base of a single ommatidium terminate at different levels of the visual pathway. The six "short" axons (1-6) end within the lamina ganglionaris; the two "long" axons stemming from cells 7 and 8, bypass the lamina and terminate in the medulla (6,7,14,19).

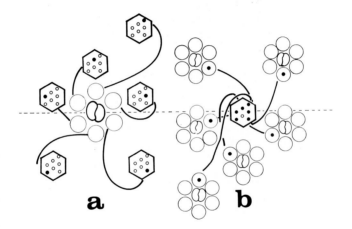

FIG. 1. Schematic representation of the main structural characteristics of the dipteran eye. Axons arising from six photoreceptors "looking" at the same point in space (black dots in scheme a) are grouped together in the same synaptic unit. On the other hand, the six axons stemming from a single ommatidium are distributed in six synaptic units (scheme b). Note the trapezoidal patterns formed by the rhabdomeres, the ommatidia and the synaptic units.

4. The neural projection of the photoreceptors on the lamina follows strict geometric rules. The six "short" axons form part of six different synaptic units or "optical cartridges", distributed in the lamina in a trapezoidal pattern. In turn, the six short axons integrating a single cartridge arise from six photoreceptor cells lying within different ommatidia. These also form a trapezoidal pattern (3,4,20).

5. Each compound eye is divided into two halves. In the dorsal one, the rhabdomeric pattern is the mirror image of that appearing in the ventral half (9). At equator level, ommatidia exhibiting enantiomorphic

rhabdomeric patterns face each other along a zigzag line.

Keeping in mind this overview of the mature dipteran eye, let us turn our attention to the developmental processes involved in cell differentiation, proper organization of the neural connections, and cell patterning.

FROM AN EPITHELIAL CELL TO A BIPOLAR NEURON

The photoreceptors of muscoid flies are derived from epithelial cells forming the so-called imaginal eye-disks. These are special larval structures composed of cell populations set apart from the cell lines giving origin to larval tissues during embryogenesis (15). Most typical organs of the adult insect, such as wings, legs, genitals and the main sense organs, arise from different imaginal disks. The developmental process begins at early larval stages, but the most dramatic morphological changes occur during a special single instar interposed between the last larval instar and the adult. This instar is the pupa. Consequently, analysis of the morphogenesis of photoreceptor cells must cover both the motile larval period and the quiescent pupal stage.

During the first and second larval instars all the imaginal disk cells look alike. They appear as cuboidal elements bearing scarce, short,

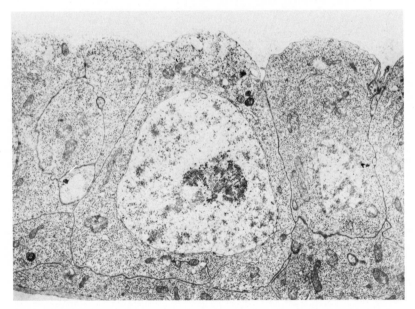

FIG. 2. At early larval stages the future photoreceptors appear as cuboidal epithelial cells particularly rich in ribonucleoprotein particles. Electronmicrograph of a second-instar disk of <u>Phaenicia</u>. (X 5950)

irregular microvilli (Fig. 2). The cytoplasm contains a large number of ribonucleic protein particles, most as monosomes. As recognized in other epithelia, the disk cells are interconnected by a series of junctions. At these very early stages there are, starting at the cell surface: a zonula adhaerens, septate desmosomes and gap junctions.

Desmosomes have also been seen close to the basal lamina. In third-instar disks the undifferentiated cells are columnar. The cytoplasm shows a few cisternae of endoplasmic reticulum occurring in the perinuclear region. Running longitudinally in respect to the main axis of the cell, there are conspicuous bundles of tubules that appear concentrated near the apical pole.

Classic as well as modern investigations (1,21) indicate that differentiation of photoreceptor cells begins at the middle of the third-instar period (50-60 h after hatching). The first clusters of neuron-like elements appear at the caudal pole of the disk (Fig. 3).

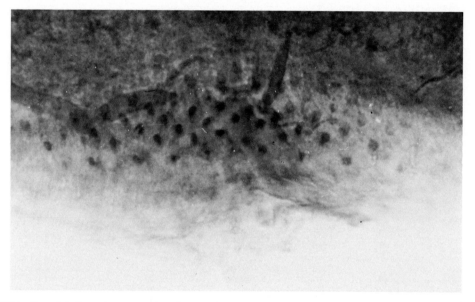

FIG. 3. Light micrograph of the first group of protoommatidia appearing in a third-instar disk approximately 50 h after hatching. Lead sulfide stain, <u>Phaenicia</u>, X 250.

Concomitantly, a transverse groove or furrow (17,23) on the surface of the disk delineates a clear separation between the still undifferentiated cells of the anterior zone (anterior cellular field, ACF, 23) and the protophotoreceptor clusters (or protoommatidia) lying close to the optic stalk.

Golgi stained, the early differentiated cells appear as typical bipolar neurons (Fig. 4). In some of them, the apical dendrite is extremely short whereas in others it exhibits a more conventional appearance. Electron microscopic studies have demonstrated that both cell types coexist in the same protoommatidium. Axons arising from these neural elements cross the disk basal lamina and penetrate the optic stalk.

CELL PATTERNING IN THE DISK

Transverse sections of the imaginal epithelium illustrate the spatial arrangement of protophotoreceptors. Each cluster is composed of eight

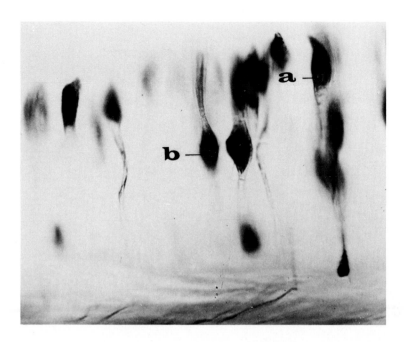

FIG. 4. Light micrograph of the general morphology of early differentiated protophotoreceptors. Two main types of cells (a and b) can be
distinguished. Type b corresponds to protophotoreceptor 8. Golgi stain,
Phaenicia. (X 1425)

cells (Fig. 5). Near the epithelial surface, cells 1,2,5,6 and 8 are
represented by relatively small areas of cytoplasm, conversely cells
3,4 and 7 exhibit their maximum size. It is important to note that
cell 8 bears a long apical dendrite; therefore, its soma is deeply
located with respect to the epithelial surface. At the level of the
soma of this central cell, other components of the protoommatidium
appear to form flat slender processes. A similar cell configuration
has been observed by Ready et al. (17) in the eye-disk of Drosophila.
These investigators also found, "immediately posterior to the furrow",
preclusters of only five cells. Based on autoradiographic and mosaic
inducing techniques, they were able to demonstrate that these cells will
become photoreceptors 2,3,4 and 8 of the mature ommatidium. The final
number of photoreceptors is completed by recruitment of three additional cells from surrounding undifferentiated elements. According to
Ready et al. (17) the formation of the eye pattern is not dependent
on cell lineages. This point of view is also supported by the grafting
experiments of Shelton and Lawrence (18) on Oncopeltus that indicated
that the cells forming an ommatidium "are normally derived from several
unrelated cells".
It is important to emphasize that the spatial arrangement of the pro-
toommatidial units follows a consistent geometrical pattern. Ommatidial
precursors in the dorsal half of the disk exhibit cell arrangements
that are the mirror images of those appearing in the ventral half (17,
24 and Fig. 6). The enantiomorphic units meet at equatorial level, as
occurs in the mature eye. In the eye-disk, however, the precise inter-

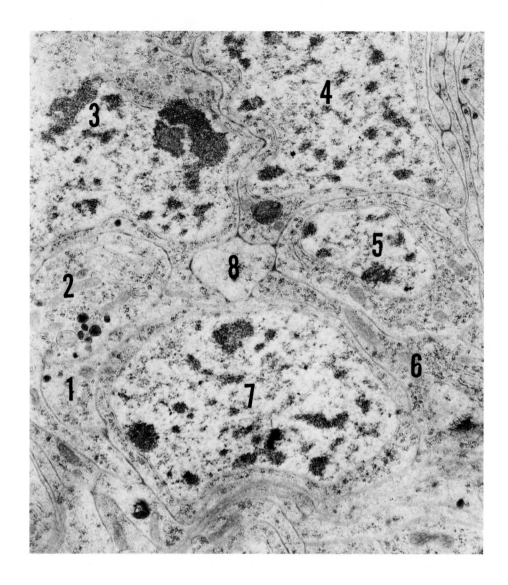

FIG. 5. Cross section through the apical region of a protoommatidium of <u>Phaenicia</u>. Note seven peripheral protophotoreceptors (1-7) and a central one (8). X 15,000.

locking of the units is lacking. As already reported (24) we have observed in this region a three to one or a two to one alternation of specular protoommatidia. In contrast with the precocious order found in other parts of the disk, the equatorial area seems to reach complete maturity at later developmental stages.

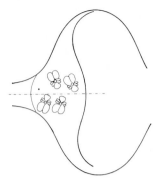

FIG. 6. Schematic drawing illustrating the
distribution of two kinds of enantiomorphic
protoommatidia occurring in the third-instar
disks of muscoid flies.

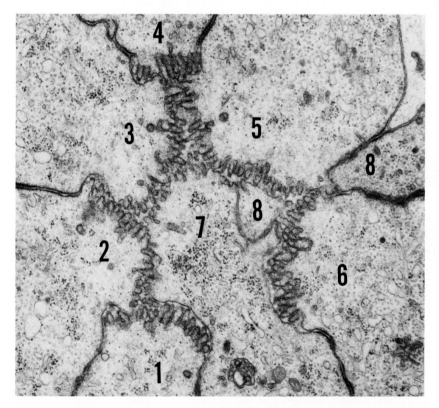

FIG. 7. Electronmicrograph of a cross section through a protoommatidium
from a 48 h pupa. 1-6 peripheral photoreceptors; 7-8 superior and
inferior central cells. (X 11700)

DEVELOPMENT OF THE RHABDOMERES

During the pupal stage the protophotoreceptors undergo important morphological changes. These involve a dramatic shortening of the cell
bodies and the parallel formation of plasma membrane folds along the
cell surfaces facing the center of the protoommatidium (16,21,25). As
maturation proceeds, these folds become more regular and numerous, inter-
digitating with those arising from opposing cells.

As shown in Fig. 7, 48 h after puparium formation, the immature
rhabdomeric microvilli form a closed-type of central rhabdom. Serial
section analysis demonstrates that protophotoreceptor 8 also participates
in the formation of this primitive rhabdom.

Approximately 90 h after pupation the central ommatidial cavity
begins to appear. At this stage each photoreceptor bears a well developed rhabdomere consisting of thousands of highly ordered microvilli.
The photoreceptors increase their length in the later part of the pupal
period.

MATURATION OF THE SYNAPTIC POLE

So far, I have focused upon the geometric arrangement of cell bodies
and the differentiation of photoreceptive structures. However, the most
characteristic properties of the dipteran eye are associated with
neural connections linking the photoreceptors to the second order neurons
(11). Let us now trace the pathway and axon terminal of each protophotoreceptor cell. The axons penetrate the optic stalk, which connect
each eye-disk to the homolateral cephalic lobe. During the first hours
of larval life the optic stalk is mainly a glial organ containing a very

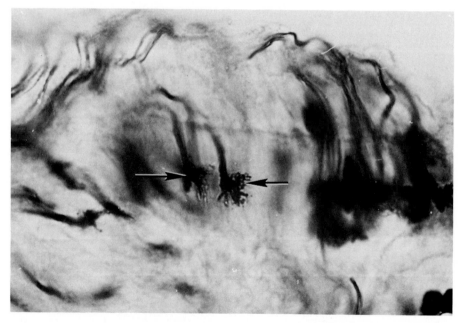

FIG. 8. This light micrograph shows numerous filopodia (arrows) arising
from the protophotoreceptor endings. Golgi stain, 40 h pupa. X 1,500.

thin nerve bundle (in Phaenicia it consists of 40 ± 7 axons). These nerve fibers do not originate in the disk, but arise, according to Bolwig (2), from a small photoreceptor organ situated close to the bucal armature of the larva.

A few hours after the beginning of the third larval period, there is a sudden increase in the number of axons running within the optic stalks (21). This new population of nerve fibers is derived from disk cells undergoing neural differentiation. As development proceeds, new axons enter into the stalks in such a way that at the prepupal stage there are more than 5000 fibers running toward each cephalic lobe.

When the optic stalk is examined in some detail it becomes evident that the 5000 fibers do not form a unitary bundle; instead they are divided into hundreds of small but distinct fascicles composed of eight axons. Where do these axons terminate? How do they terminate?

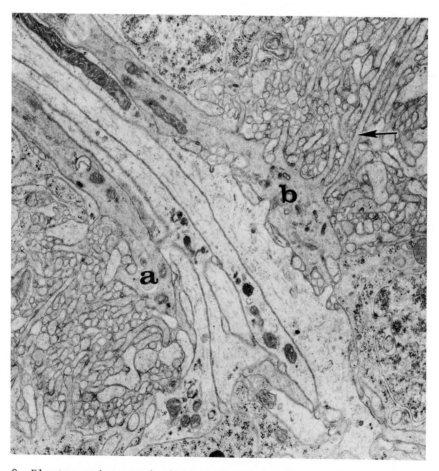

FIG. 9. Electron micrograph showing terminal segments of two protophoto-receptor axons (a,b). Note the presence of a long filopodium (arrow) arising from one of the enlarged terminals. The clear fibers running between the nerve terminals correspond to the main prolongation of the second order neurons. Phaenicia, 48 h pupa. (X 16200)

These two relevant questions were answered by a single fortunate Golgi-stained section obtained in 1971 that showed that the axons arising from protophotoreceptors lying in the disk terminate as a palisade in the peripheral regions of the homolateral cephalic lobe. The nerve endings in Golgi-stained material appear as oval enlargements covered by numerous tiny processes (Fig. 8). Electron microscopy has provided complementary information that clearly demonstrates that these tiny processes are actually typical filopodia, analogous to those occurring in the terminal cones of growing nerve fibers.

As reported by Meinertzhagen (13) the filopods form in mid third-instar and late third-instar larvae as a dense plexus at the level of the lamina-anlage. Taking advantage of the resolution offered by the electron microscope and following serially the group of axons stemming from the base of a single protoommatidium, it has been possible to demonstrate that at larval stages and during the first hours of the pupal period, they terminate together in one restricted area of the lamina-anlage (21). The retina-lamina projection pattern is, therefore,

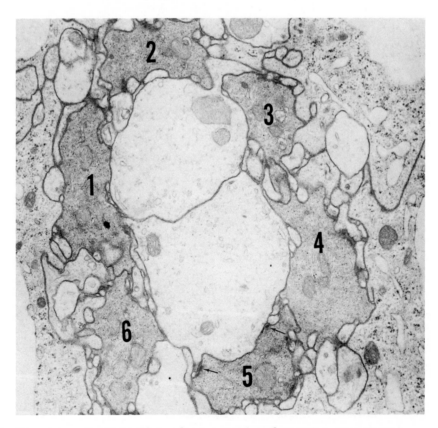

FIG. 10. Approximately 90 h after puparium formation transverse sections passing through the lamina-anlage reveal the presence of typical optical cartridges. At this developmental stage the photoreceptor endings (1-6) lack "capitate projections". Arrows indicate two immature synaptic junctions. <u>Phaenicia.</u> (X 16200)

of the uncrossed type similar to that found in insects with a closed
central rhabdom.

The protophotoreceptor endings retain filopodia developed during
larval periods; at later maturation stages these processes appear clearly
polarized, arising exclusively from one side of the terminal cone (Fig.9).
Membrane densifications occur at sites where neighbor axons meet.   It
is important to note that we have never observed at this stage T-shaped
synaptic bars or accumulations of synaptic vesicles.

The adult topological relationship between the retina and the lamina
is initiated by outgrowth of a single thick process from the terminal
cone. This thick branch follows a parallel course with respect to the
surface of the lamina-anlage. Ninety hours after puparium formation
the distal part of the branch changes its primitive direction and
runs toward the brain centers following the main prolongations of the
second-order neurons; electron microscopy shows typical synaptic units
(optical cartridges, Fig. 10). This means that the adult-type of
projection pattern is now fully developed. The specializations at the
level of the synaptic loci appear 120-140 hours after pupation. However, the "capitate projections" (invaginations of the glial cells within minute infoldings of the axon membrane) develop later, close to the
emergence of the imago.

IS A FIXED GEOMETRICAL PATTERN PREDETERMINED IN EYE-DISK CELLS?

The distribution of the future photoreceptors is not random but
follows a consistent geometrical pattern. Consequently, the third-
instar disk contains two enantiomorphic populations of protoommatidia
separated by an irregular equatorial line. Each disk also contains an
important number of still undifferentiated, unpatterned cells. This
means that when a disk is transferred from a donor to a host larva,
these undifferentiated cells will continue their developmental process
within the host. Such a peculiar situation allows us to test alternative hypotheses concerning the potential geometrical plasticity of the
system by using dissimilar kinds of disk fragments.

An implant consisting of a homogeneous population of protoommatidia
and an undetermined number of undifferentiated cells could give rise to
two kinds of heterotopic eyes. If the undifferentiated cells are already
determined to follow the normal ventral or dorsal aggregation pattern,
the resulting eyes will show a uniform, single type of ommatidia.
Conversely, if there is some sort of "pattern regulation" the heterotopic eye will exhibit an equatorial line and the two normal types of
specular ommatidia. From a total of 46 eyes derived from implants of
ventral or dorsal disk fragments, 10 were serially cut and system-
atically photographed in order to obtain complete retinal maps. After
a careful study of 2470 ommatidia, only 46 (less than 2%) showed
enantiomorphic rhabdomeric patterns. On the other hand, a study of 10
eyes derived from fragments containing the medial region of the disk
consistently revealed the two ommatidial patterns occurring in normal
eyes. These results strongly suggest that the mechanism of cell aggregation is not endowed with any degree of geometrical plasticity (24).

We have also performed a third type of experiment in which fragments
containing exclusively medial parts of the ACFs (i.e., a uniform
population of undifferentiated cells) were transferred to host larvae.
The resulting eyes showed the two specular kinds of ommatidia. It
is important to note that these results are difficult to interpret

within the conceptual framework offered by the "template hypothesis" as
it was advanced by Ready et al. (17). In our experiments one of the
planes of section separates the already formed ommatidial precursors,
which may act as templates, from the still undifferentiated cells. The
different types of transplantation experiments are summarized in Fig. 11.
     Considering that the heterotopic eyes are neurally isolated sense
organs (10), they appear to be a suitable material for assessing the
role played by the "target neurons" during the development of the nerve
connections. Our investigations (22,24) indicate that the transplanted
eyes lack the so-called external chiasm (6). The photoreceptor axons
aggregate at random, forming a pseudolamina in which the normal synaptic
modules are absent. At least in this instance the presence of second
order neurons seems to be essential for the development of a normal
spatial distribution of the photoreceptor axons.

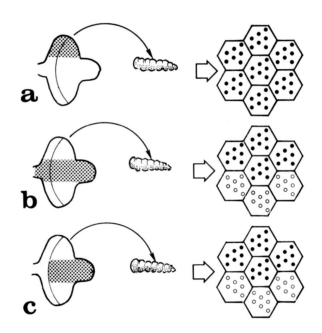

FIG. 11. When a dorsal (or ventral) eye-disk fragment is transferred
to a host larva, the resulting heterotopic eye shows a single population
of ommatidia (a). Conversely, the heterotopic eyes derived from medial-
disk parts exhibit the two normal, enantiomorphic ommatidial populations (b). These two specular kinds of ommatidia also occur in heterotopic eyes derived from fragments containing, exclusively, undifferentiated cells (c).

REFERENCES

  1. Bodestein,D.(1950): In: <u>Biology of Drosophila</u>, edited by M.Demerec,

pp. 275-367. Wiley, New York.

2. Bolwig,N.(1946): Vidensk. Medd. fra Dansk naturh. Foren., 109:81-217.

3. Braitenberg,V.(1966): Z. Verlg. Physiol., 52:212-214.

4. Braitenberg,V.(1967): Exp. Brain Res., 3:271-298.

5. Cajal,S.R.(1909): Trab. Lab. Invest. Biol. (Madrid)., 7:217-257.

6. Cajal,S.R., and Sánchez,D.(1915): Trab. Lab. Invest. Biol. (Madrid)., 13:1-164.

7. Campos-Ortega,J.A., and Strausfeld,N.J.(1972): Z. Zellforsch. Mikrosk. Anat., 124:561-585.

8. De Vries,H.(1956): Prog. Biophys. Mol. Biol., 6:207-264.

9. Dietrich,W.(1909): Z. Wiss. Zool., 92:465-539.

10. Eichembaun,D., and Goldsmith,T.(1968): J. Exp. Zool., 169:15-32.

11. Kirschfeld,K.(1967): Exp. Brain Res., 3:248-270.

12. Kuiper,J.(1962): Symp. Soc. Exp. Biol., 16:58-71.

13. Meinertzhagen,I.(1973): In: Developmental Neurobiology of Arthropods, edited by D. Young, pp. 51-104. Cambridge University Press, London.

14. Melamed,J., and Trujillo-Cenóz,O.(1968): J. Ultrastruct. Res., 21:313-334.

15. Nothinger,R.(1972): In: The Biology of Imaginal Disks, edited by H. Ursprung and R. Nothinger, pp. 1-34. Springer-Verlag, New York, Heidelberg, Berlin.

16. Perry,M.(1968): J. Morphol., 124:227-248.

17. Ready,D.,Hanson,T., and Benzer,S.(1976): Dev. Biol., 53:217-240.

18. Shelton,P., and Lawrence,P.(1974): J. Embryol. Exp. Morphol., 32: 337-353.

19. Trujillo-Cenóz,O.(1965): J. Ultrastruct. Res., 13:1-33.

20. Trujillo-Cenóz,O., and Melamed,J.(1966): J. Ultrastruct. Res., 16:395-398.

21. Trujillo-Cenóz,O., and Melamed,J.(1973): J. Ultrastruct. Res., 42:554-581.

22. Trujillo-Cenóz,O., and Melamed,J.(1975): Naturwissenschaften. 62:42.

23. Trujillo-Cenóz,O., and Melamed,J.(1975): J. Ultrastruct. Res., 51:79-93.

24. Trujillo-Cenóz,O., and Melamed,J.(1978): <u>J. Ultrastruct. Res.</u>, 64:46-62.

25. Waddington,C., and Perry,M.(1960): <u>Proc. R. Soc. Lond. (Biol)</u>, 153:552-576.

Visual Cells in Evolution, edited by Jane A. Westfall, Raven Press, New York © 1982.

The Appearance of Specific Neuronal Properties During the Development of Photoreceptors and Horizontal Cells in *Xenopus Laevis*

Joe G. Hollyfield, Jeanne M. Frederick, Dominic Man-Kit Lam, and Mary E. Rayborn

Cullen Eye Institute, Program in Neuroscience, Baylor College of Medicine, Houston, Texas 77030

Studies of differentiation in the nervous system using morphological techniques are usually limited to descriptions of the time and appearance of various synaptic contacts and the development of specific neuronal shapes. Biochemical studies have documented the times at which putative neurotransmitters and their enzyme systems appear and in some instances the times of appearance of high affinity uptake mechanisms for neurotransmitters or their precursors. Usually these studies are independent investigations; rarely are attempts made to correlate cytological differentiation of neurons with the appearance of specific biochemical and physiological parameters. Through the development of autoradiographic techniques for visualizing cells with specific high affinity uptake properties, it is possible to document the times of appearance of these properties when the cells of the developing neuroepithelium show no morphologically distinctive features. We have initiated a series of studies to evaluate the appearance and maturation of specific neuronal properties during the development of the retina of Xenopus laevis using a combination of autoradiographic, physiological and biochemical techniques. The primary thrust of these investigations is to determine the interrelationship between the onset of neurotransmitter synthesis, mechanisms for the release of neurotransmitters and the system for inactivation of the transmitters by high affinity reuptake mechanisms.

Because this symposium is focused on the photoreceptor, the studies described in this paper will deal with the appearance of the high affinity uptake mechanisms for the free amino acid, taurine, during photoreceptor differentiation. Although taurine does not appear to be a neurotransmitter in the retina (2), its high concentration in the photoreceptors along with the known degenerative changes following the depletion of taurine in animals that cannot synthesize taurine make the study of the uptake mechanisms for this compound by photoreceptors an important area of investigation. In addition, results will be summarized on the development of the γ-aminobutyric acid (GABA) properties in the horizontal cell of Xenopus. Considerable evidence implicates GABA as a neurotransmitter in this cell type. GABA-ergic horizontal cells form

synaptic connections with both rod and cone photoreceptors in <u>Xenopus</u>
retina. The time of appearance of the high affinity uptake, synthesis
and release of GABA in relationship to the time of synapse formation between these cell types will be discussed.

METHODS

Laboratory maintained <u>Xenopus laevis</u> were the source of all embryos
used in this study. Embryos were staged according to Nieuwkoop and
Farber (7). Autoradiographic evaluations of the high affinity uptake
systems were based on the ability of the glutaraldehyde fixative to
crosslink compounds containing free amino groups in cells. Retinas were
dissected from embryos at stages 27 through 45 and from stage 55 in
juvenile frogs. They were incubated for 10 to 30 minutes with either
^{3}H-GABA (100 μCi/ml, 44 Ci/mM) or ^{3}H-taurine (50 μCi/ml, 23.0 Ci/mM).
Retinas were incubated in and washed with double strength Niu-Twitty
solution and fixed on ice for 10 minutes in 2% glutaraldehyde in 50 mM
phosphate buffer at pH 7.2, then postfixed in 1% OsO_4 for an additional
20 minutes. Sections were cut following embedment in Epon at 1 μm
thickness and processed for light microscope autoradiography according to
previously described methods by Hollyfield et al. (5). GABA synthesis
was determined by incubating freshly dissected eyes with ^{3}H-labeled glutamic acid followed by transmitter separation of products using high
voltage electrophoresis as described by Hildebrand et al. (4). The study
of neurotransmitter release properties utilized the procedure of Sarthy
and Lam (9). Ten to twenty eyes were incubated for 15 minutes in either
^{3}H-taurine or ^{3}H-GABA. After several rinses in fresh media, two eyes
were removed for determination of radioactivity and the rest were transferred to graduated conical tubes containing 10 mls of media. After 10
minutes the media was replaced with 2 ml of fresh media. Aliquots of
1.9 ml were withdrawn at two minute intervals, mixed with aquasol and the
radioactivity determined. Fresh medium was added to make up a 2 ml
volume. At appropriate times the eyes wre exposed to Niu-Twitty medium
containing 56 mM K^+. Radioactivity in retinas was counted at both the
beginning and end of the release experiment.

RESULTS

Taurine and the Development of Photoreceptors

Retinal rudiments at the earliest developmental stages surveyed (stage
26) consist of cells in a typical neuroepithelial configuration containing numerous yolk platelets and lipid droplets. In autoradiographs of
these early stages, following incubation with ^{3}H-taurine, silver grains
were distributed diffusely over the tissue with no indication of preferential uptake by any neuroepithelial cells. At stage 35/36 when
nuclear stratification segregated the perspective photoreceptors to the
outer layer of the developing retina, some photoreceptors acquired
specific uptake channels for ^{3}H-taurine as evidenced by the localization
of silver grains over some photoreceptor cells in this region (Fig. 1).
Usually these cells are situated in the middle of the retinal rudiments
adjacent to the pigment epithelium. By stage 37/38 (Fig. 2) short outer
segments first begin to appear on some of the differentiating photoreceptors, and at this stage increasing numbers of photoreceptors show
^{3}H-taurine specific labeling. Although it is not possible at early stages

to determine precisely whether these photoreceptors are rod or cone types, by stage 40 (Fig. 3) the two populations of photoreceptors can be easily distinguished, and ^{3}H-taurine specific labeling of cone photoreceptors is evident. Preferential labeling of cone photoreceptors continues through stage 44, and at this time some cells in the inner nuclear layer begin to accumulate ^{3}H-taurine as well. In retinas from mid-larval stage tadpoles and juvenile frogs, both rod and cone photoreceptors take up ^{3}H-taurine although cone labeling is more intense than rod labeling when retinas are incubated in the light. Also, in the tadpole and juvenile retinas a cell type in the inner nuclear layer, probably a type of amacrine cell, shows specific taurine labeling. It should be pointed out that the horizontal cells in this retina do not take up ^{3}H-taurine.

As reported in detail by Frederick et al. (2), the photoreceptor cells show high affinity uptake mechanisms for taurine, but these cells are unable to release ^{3}H-taurine in response to stimuli that readily promote release of other putative transmitters. When double label experiments using ^{14}C-labeled-taurine and ^{3}H-conventional transmitters, utilized by neurons in the inner nuclear layer (GABA, glycine, or dopamine) are performed and the retinas then subjected to high K$^+$ concentration, large amounts of the ^{3}H-labeled transmitters are released, whereas ^{14}C-taurine is not released above baseline efflux levels. In addition, amino acid

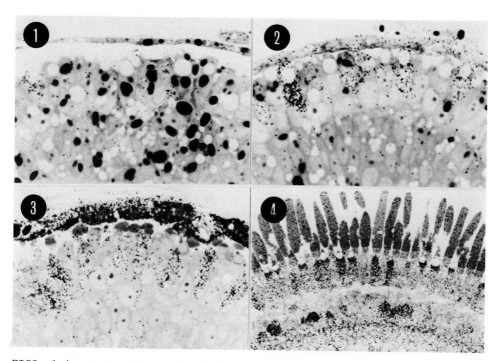

FIGS. 1-4. Autoradiographs of <u>Xenopus laevis</u> retinas at various stages of development incubated in the presence of ^{3}H-taurine. Notice the progressive increase in number of developing photoreceptors with specific uptake of this amino acid. 1: Stage 35/36; 2: Stage 37/38; 3: Stage 40; 4: Juvenile frog retina. See text for further explanation.

analysis of the developing Xenopus retina indicates that measurable a-
mounts of taurine are not present until after stage 46. In contrast,
glycine and GABA can be detected as early as stage 38/39. Since light-
evoked responses can be recorded from developing Xenopus retinas as
early as stage 38/39 and the full component of the ERG is present from
stage 40 onward (11), the absence of taurine during the stages when photoreceptors are first responsive to light indicates that taurine is not
a critical metabolite for photoreceptor differentiation. Nevertheless,
studies with adult animals that are unable to synthesize taurine show
profound degenerative changes in the photoreceptor cells when this amino
acid is depleted from the diet. Although our studies indicate that taurine is not required for differentiation of the photoreceptors, it appears to play some critical role in the maintenance of photoreceptors
in adults (3).

γ-Aminobutyric Acid and the Development of Horizontal Cells

Horizontal cells of the vertebrate retina have direct synaptic contact with the bases of photoreceptor cells. This synaptic specialization in Xenopus has been studied in detail by Chen and Witkovsky (1)
and by Witkovsky and Powell (12). We have demonstrated also that the
horizontal cells in Xenopus have specific uptake properties for ^{3}H-GABA
(6). Furthermore, horizontal cell processes that label with ^{3}H-GABA in
Xenopus contact both rod and cone photoreceptor cell types, suggesting
the presence of only one type of horizontal cell in this species. We
have followed the time of appearance of high affinity uptake systems in
Xenopus retina and have found that prior to stage 31 retinas incubated
in the presence of ^{3}H-GABA show only diffuse labeling over the somata
of the neuroepithelial cells (Fig. 5). At stage 32 however, a small
population of the neuroepithelial cells show preferential uptake of ^{3}H-
GABA (Fig. 6). At this stage stratification of retinal neurons has not
commenced. By stage 35/36 when retinal histogenesis has begun, the cells
that show specific ^{3}H-GABA uptake are clustered along the border of the
developing outer plexiform layer (Fig. 7) in the location where horizontal cells reside in the adult retina. A characteristic feature of all
late tadpole and adult retinas incubated in ^{3}H-GABA is the continuous
band of radioactivity over horizontal cells (Fig. 8).
GABA synthesis can first be detected in the Xenopus retina at around
stage 35/36, and the rate of synthesis increases steadily until stage
44/45 when it approaches 80 percent of the rate measured in adult retinas. To follow the time of development of release properties of the
horizontal cell, retinas preloaded with ^{3}H-GABA, beginning at stage 32
onward were subjected to high K^+ concentration. Although GABA can be
taken up as early as stage 32 of development, no detectable release was
observed until stages 37/38. In order to understand the relationship
between the development of the release properties of the horizontal
cell and the appearance of horizontal cell synapses, we have studied
the maturation of the synaptic complex at the photoreceptor base during
these critical stages. The synaptic lamellae of the photoreceptor cells
are not present prior to stage 35/36. When the synaptic lamellae are
first detected, they can be observed free within the cytoplasm near the
photoreceptor base or adjacent to the presynaptic membrane (Figs. 9-10).
Usually at these early stages only a few synaptic vesicles are associated
with the synaptic lamellae. The contour of the presynaptic membrane is
usually straight with no indication of the characteristic invaginations

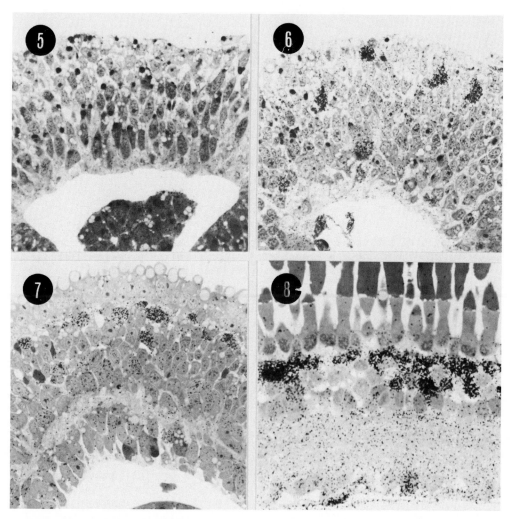

FIGS. 5-8. <u>Xenopus laevis</u> retinas at different stages of development illustrate the appearance of uptake systems for ^{3}H-GABA in horizontal cells. 5: Stage 30 with no specific uptake of ^{3}H-GABA; 6: Stage 31 has several cells with specific ^{3}H-GABA uptake before histogenesis has commenced; 7: Stage 35/36 with ^{3}H-GABA labeled cells situated in the position where horizontal cells typically reside; 8: ^{3}H-GABA labeled cells in a juvenile frog retina for comparison with Fig. 7.

filled with postsynaptic processes present in mature photoreceptors. Around stage 37/38 postsynaptic processes, presumably of horizontal and bipolar origin, extend into deep invaginations in the photoreceptor base (Figs. 11 and 12). Increased membrane densities are now present on both the pre- and postsynaptic elements. By stage 41 the synaptic complexes at the bases of both rod (Fig. 13) and cone (Fig. 14) photoreceptors show all the features present in late tadpole and juvenile frog retinas.

The time of development of synaptic specializations between photoreceptor and horizontal cells correlates precisely with the time when
the horizontal cells are capable of releasing their putative neurotransmitters. Thus, the synaptic contacts between horizontal cells and
photoreceptor are likely sites of GABA release. [3]GABA can be taken
up by the developing horizontal cells as early as stage 32 but cannot
be released from the horizontal cells until the appearance of morphologically distinct synaptic contacts. This strongly implies that the
channels for reuptake of ^{3}H-GABA by the horizontal cell are distinctly
different from the release channel. The former are specialized membrane
channels that cannot be distinguished by the morphological tools utilized in this study, whereas the latter are membrane specializations
identifiable as conventional synapses between the horizontal cells and
the photoreceptor.

CONCLUSION

Taurine uptake by cone photoreceptors first occurs at a stage immediately preceding the appearance of small tips of outer segment membrane.
Although we have measured by amino acid analysis high concentrations
of endogenous taurine in the adult Xenopus retina, taurine is not detectable in the developing retina at stages when photoreceptor outer segments are first elaborated and light-evoked responses can first be recorded. The absence of endogenous taurine during these critical stages
of photoreceptor development indicates that this metabolite is not required for the early phases of photoreceptor development and physiological differentiation. The importance of retinal taurine for maintenance of adult photoreceptors remains to be defined, and use of a specific taurine analogue may facilitate this analysis.
The coupling of high affinity uptake properties of neurons with auto-
radiograhic techniques provides a new dimension for studies of neuronal
differentiation. By morphological techniques alone specific neuronal
cell types cannot be identified until specific synaptic connections or
cell shapes develop. The early appearance of the high affinity uptake
sites for ^{3}H-GABA in horizontal cells allows us to distinguish this
cell type several stages before morphologically distinct features appear.
This aspect of neuronal differentiation is a useful early marker in the
identification and characterization of development of other neuronal
cells in the retina as shown by Rayborn et al. (8) and Sarthy et al.
(10).

FIGS. 9-14. Development of synaptic complexes at the bases of photoreceptors in Xenopus laevis. 9-10: Stage 35/36 with synaptic lamellae
adjacent to the plasma membrane at the bases of the photoreceptors.
Invaginations of postsynaptic processes are not evident. 11-12: Stage
37/38. Invaginating processes in bases of photoreceptors are evident;
synaptic lamellae appear longer and are surrounded by more synaptic
vesicles. 13-14: Stage 41 with adult features of photoreceptor
synapses on second-order neurons.

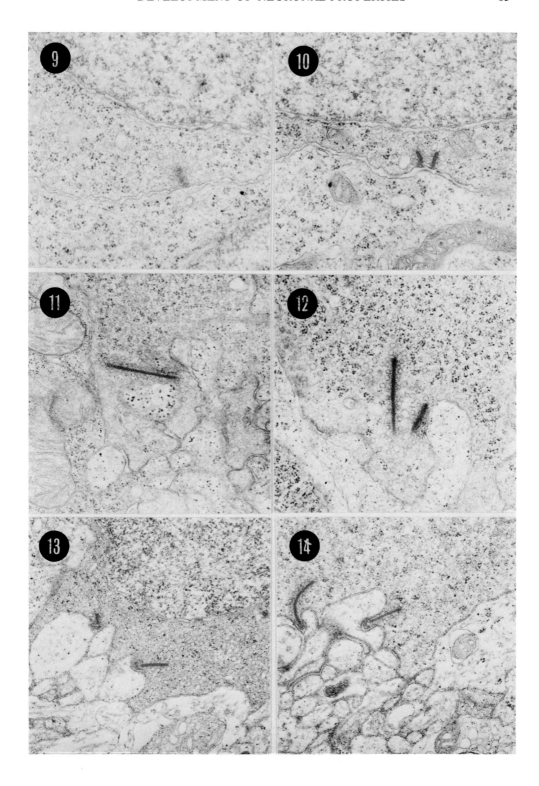

REFERENCES

1. Chen,F., and Witkovsky,P.(1979): J. Neurocytol., 7:721-740.

2. Frederick,J.M.,Lam,D.M.K., and Hollyfield,J.G.(1980): In: The Structure of the Eye, edited by Joe G. Hollyfield, Elsevier/North Holland, Inc., New York. (in press).

3. Hayes,K.C.,Rabin,A.R., and Berson,E.L.(1975): Am. J. Pathol., 78:505-524.

4. Hildebrand,J.G.,Barker,D.L.,Herbert,E., and Kravitz,E.A.(1971): J. Neurol. Biol., 2:231-246.

5. Hollyfield,J.G.,Besharse,J.C., and Rayborn,M.E.(1977): J. Cell Biol., 75:490-506.

6. Hollyfield,J.G.,Rayborn,M.E.,Sarthy,P.V., and Lam,D.M.K.(1979): J. Comp. Neurol., 188:587-598.

7. Nieuwkoop,P.D., and Farber,J.(1956): Normal Tables of Xenopus laevis (Daudin). North Holland Publishing Co., Amsterdam.

8. Rayborn,M.E.,Sarthy,P.V.,Lam,D.M.K., and Hollyfield,J.G.(1980): J. Comp. Neurol., 195:585-594.

9. Sarthy,P.V., and Lam,D.M.K.(1979): J. Neurochem., 32:1269-1277.

10. Sarthy,P.V.,Rayborn,M.E.,Hollyfield,J.G., and Lam,D.M.K.(1980); J. Comp. Neurol., 195:595-602.

11. Witkovsky,P.,Gallen,E.,Hollyfield,J.G.,Ripps,H., and Bridges,C.D.B. (1976): J. Neurophysiol., 39:1272-1287.

12. Witkovsky,P., and Powell,C.C.(1981): Proc. R. Soc. Lond. (Biol), 211:373-389.

Visual Cells in Evolution, edited by Jane A. Westfall,
Raven Press, New York © 1982.

Identification of Neurotransmitter Candidates in Invertebrate and Vertebrate Photoreceptors

Dominic Man-Kit Lam, Jeanne M. Frederick, Joe G. Hollyfield, P. Vijay Sarthy, and *Robert E. Marc

*Cullen Eye Institute, Program in Neuroscience, Baylor College of Medicine, Houston, Texas 77030; *Sensory Sciences Center, Department of Ophthalmology, University of Texas Health Science Center, Houston, Texas 77025*

Anatomical and physiological studies have established that in most invertebrate and vertebrate visual systems, synaptic transmission from photoreceptors to second order neurons is chemically mediated. Identities of the neurotransmitters used by photoreceptors, however, are not known in any retina. This article is a survey of the candidates that have been suggested as photoreceptor neurotransmitters in a variety of invertebrate and vertebrate retinas.

METHODS

The procedures used for studies presented in this article have all been described in detail elsewhere (e.g., 15,22). All retinal preparations were done in vitro.

RESULTS

Cephalopod Photoreceptors

As a first example to identify possible neurotransmitters used by invertebrate photoreceptors, we have chosen to study transmitter syntheses by cephalopod retinas. The cephalopod retina is a relatively simple structure consisting of non-neural supporting epithelial cells and two neural elements: photoreceptors and efferent nerve fibers (4,12,20,37). The photoreceptor axons project to the optic ganglion but, before leaving the retina, they give off collaterals to the retinal plexus. Efferent fibers, originating from cells in the optic ganglion, also terminate in the plexus. There are junctions within the plexus having the ultrastructural appearance typical of chemical synapses (4,12). We have attempted to identify the neurotransmitters in the cephalopod retina using a screening procedure of Hildebrand et al. (13). Neural tissue was incubated with radioactive precursors of transmitter candidates and the subsequent synthesis and accumulation of transmitters were analyzed biochemically.

Four species of cephalopods, Octopus joubini (5-20 g), Octopus vulgaris (200-500g), Lolliguncula brevis (dwarf squid, 10-20 g) and Loligo pealei (common squid from Woods Hole, 30-80 g) were used in this study. The animals were anesthetized with 2% ethanol in sea water. The eyes alone, or eyes connected to optic ganglia by optic nerves were excised and dissected free of cornea, lens and other tissues. The eyes and ganglia were incubated at 23°C for 1-4 h in an oxygenated Leibovitz medium (L-15, about 200 mg tissue/ml), supplemented with salts, glucose, fetal calf serum and 20 µCi/ml of one or a combination of methyl-(^{14}C) choline chloride (30 Ci/mole), L-(^{14}C)glutamic acid (uniformly labeled, 255 Ci/mmole) and L-(^{14}C)tyrosine (uniformly labeled, 54 Ci/mole). After incubation, the tissues were washed in unlabeled medium for 5 min and the retinas, optic nerves, and optic ganglia were separated by dissection. The tissues were then each homogenized in electrophoresis buffer (1.4 M acetic acid-0.47 M formic acid, about 200 mg tissue/ml), containing unlabeled precursor molecules and transmitter candidates for use as marker standards for electrophoresis. The homogenate was centrifuged at 100 X g for 10 min. Radioactive products in the supernatant were separated and identified by high-voltage paper electrophoresis (pH 1.9, 6000 V, 1.5 h) and ascending paper chromatography as described elsewhere (16,21). Ten microliters of the incubation medium were also analyzed for labeled products.

TABLE I

BIOSYNTHESIS OF NEUROTRANSMITTER CANDIDATES IN THE CEPHALOPOD RETINA

Retinas were incubated in L-15 medium containing ^{14}C-labeled tyrosine, glutamate and choline for 4 h. The numbers represent means ± standard deviation of radioactive counts/min (background subtracted) for 10 µl retinal homogenate for 4 experiments. Radioactivity in medium (background) was about 35 counts/min and did not show a peak in any region corresponding to the transmitter candidates.

	[^{14}C]Nor-adrenaline	[^{14}C]Dop-amine	[^{14}C]Octop-amine	[^{14}C]GABA	[^{14}C]Acetyl-choline
Octopus joubini	12 ± 10	335 ± 52*	14 ± 10	15 ± 12	8281 ± 250*
Octopus vulgaris	10 ± 8	390 ± 82*	12 ± 8	14 ± 12	10120 ± 285*
Lolliguncula brevis	10 ± 8	192 ± 42*	10 ± 8	15 ± 13	2415 ± 140*
Loligo pealei	12 ± 10	208 ± 50*	12 ± 7	12 ± 8	3451 ± 150*

* Value indicates significant synthesis.

As shown in Table 1, isolated retinas of all four species of cephalopods synthesized and accumulated significant amounts of acetylcholine and dopamine, but not γ-aminobutyric acid (GABA), noradrenaline, or octopamine. Analyses by high voltage paper electrophoresis of the extracts from an octopus retina incubated with (^{14}C) choline and (^{14}C)tyrosine show that this retina can synthesize both acetylcholine and dopamine from their precursors (Fig. 1). The identities of both acetylcholine and dopamine were verified by paper chromatography (16,21). In addition, after paper electrophoresis, the region corresponding to (^{14}C)ace-

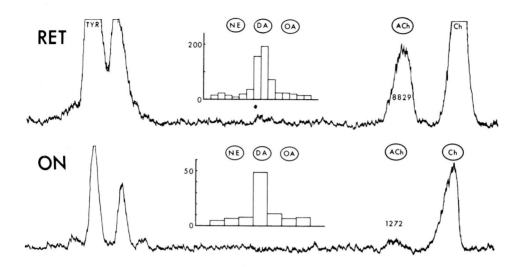

FIG. 1. Retina (RET) and optic nerve (ON) of Octopus joubini. Radio-
chemical scanning of the electropherograms of hemogenates from tissues
incubated with (^{14}C)choline and (^{14}C)tyrosine for 4 h. Distances of
migration through paper during electrophoresis: origin to DA (44 cm),
origin to ACH (63 cm). Regions corresponding to the various transmitters
were cut into strips and analyzed by scintillation counting. The amount
of radioactivity (in counts/min) associated with ACh is shown in the
scanning diagram. The NE, DA and OA activities are shown in the insert-
ed histogram. The numbers represent counts per minute above background
(about 35). TYR(tyrosine), NE(noradrenaline), DA(dopamine), OA(octopamine), ACh(acetylcholine), Ch(choline). (19).

tylcholine could be converted into (^{14}C)choline by treatment with specific acetylcholine esterase (13,30).

In experiments where the retina, optic nerve, and optic ganglion
were left connected and incubated with (^{14}C)choline and (^{14}C)tyrosine,
acetylcholine and dopamine were present not only in the retina but also
the optic nerve (Fig. 1). Furthermore, we have shown that the optic
ganglion synthesizes not only acetylcholine and dopamine, but also octopamine and noradrenaline (D.M.K. Lam, unpublished data).

The activity of choline acetyltransferase, an enzyme that synthesizes
acetylcholine, was also measured in the retina and optic nerve. The
enzymic activity was determined in tissue extracts using (^{14}C)acetate
labeled acetyl-CoA as precursor. Both the retina and the optic nerve had
significant choline acetyltransferase activities (Table 2). Again the
identity of the labeled acetylcholine was confirmed by treatment with
specific acetylcholine esterase.

Thus two transmitter candidates, acetylcholine and dopamine, were
found in the cephalopod retina, that, as mentioned above, has only two
neural elements: photoreceptors and efferent fibers. In the visual
systems of Sepia officinalis and Eledone moschata, Drukker and Schadé
(9) observed histochemically that specific acetylcholinesterase stains
were associated with the retinal plexus, and the plexiform layer in the
optic ganglion were photoreceptor terminals synapsed with second order

TABLE 2

ACTIVITY OF CHOLINE ACETYLTRANSFERASE IN THE RETINA AND OPTIC NERVE OF CEPHALOPODS

Tissues were homogenized in 9 times the wet tissue weight of 200 mM NaCl + 50 mM sodium phosphate, pH 7.1. Two μl of the homogenate was added to 10 μl of a buffer substrate containing 50 mM NaH$_2$PO$_4$, 50 mM Na$_2$HPO$_4$ (pH 7.1), 300 mM NaCl, 10 mM MgCl$_2$, 0.2 mM eserine, 10 mM choline chloride, 0.1 mM neutralized EDTA, 170 μM [^{14}C]acetyl-CoA (54.7 Ci/mole). The mixture was incubated for 15–30 min at 37 °C. Enzyme activity was linear with time (0–30 min) and tissue (1–4 μl homogenate) for this assay. Furthermore, over 90 % of the choline acetyltransferase was present in the supernatant of the homogenate.

	Octopus joubini	*Octopus vulgaris*	*Lolliguncula brevis*
Retina	2.01 ± 0.18*	2.20 ± 0.20	2.06 ± 0.18
Optic nerve	2.05 ± 0.15	2.41 ± 0.31	—**

* μmole ACh formed/h/mg wet weight ± standard deviation for 4 experiments.
** Activity not assayed.

neurons. Electron microscopic studies of octopus and squid visual systems from several laboratories have shown that photoreceptor terminals in the optic ganglion and probably photoreceptor collaterals in the retina (4,12,33) contain round, agranular synaptic vesicles similar to those found in cholinergic endings. In addition, the presence of dense-cored vesicles in presumably efferent fiber terminals has been observed (4,12,33), suggesting that these terminals may be aminergic. Our chemical results complement these histochemical and ultrastructural findings and suggest that acetylcholine is synthesized and accumulated in the photoreceptors, and dopamine in the efferent fibers.

VERTEBRATE RETINA

In the vertebrate retina, aspartic acid and glutamic acid have been suggested as candidates for photoreceptor transmitters because these compounds rapidly depolarize horizontal and 'off-center' bipolar cells from the turtle, skate and carp retinas, and also inhibit the b-wave of the electroretinogram without affecting the photoreceptor responses directly (2,8,26,27,32,35). Besides these amino acids, acetylcholine has also been suggested to be a transmitter candidate for photoreceptors. Isolated cone photoreceptors of the turtle retina synthesize and accumulate ACh (but not γ-aminobutyric acid or catecholamines) and contain some choline acetyltransferase (Acetyl-CoA: Choline O-acetyltransferase, EC 2.3.1.6) activity (15). In addition, recent physiological and autoradiographic studies suggest possible muscarinic and nicotinic receptor sites on turtle horizontal cells and goldfish bipolar cells respectively (10,31,36).

In view of these results, we have recently attempted to identify the photoreceptor neurotransmitters in different vertebrate retinas using a variety of autoradiographic, biochemical, and physiological techniques.

Turtle Photoreceptors

A direct way to identify the possible transmitters used by a neuron is to obtain a homogenous population of the cell type and analyze the transmitter candidates present and synthesized by these neurons. We have previously developed a method to dissociate the retina into single,

identified cells (15,16). By separating turtle retinal cells using
velocity sedimentation at unit gravity, a fraction containing 80 to 90%
photoreceptors can be obtained for chemical and physiological analyses.
In addition, a small number (100 to 500) of photoreceptors can be selectively drawn into a micropipet under visual observation. We have used this
method to measure the endogenous levels of aspartic acid, glutamic acid
and acetylcholine in photoreceptor-rich fractions, as well as in a small
homogeneous population of identified photoreceptors from the turtle
retina.

Free amino acids

Electrophysiological studies suggest that aspartic or glutamic acid
may be the photoreceptor transmitter. To examine the roles of these
amino acids in photoreceptor transmission, we have measured the
concentration of free aspartic acid and glutamic acid in photoreceptor-
rich fractions and in isolated retinal cells (containing all cell types)
from the turtle retina. As shown in Table 3, although dissociated cells
have a high concentration of GABA, very little was found in the photoreceptors. This is in agreement with earlier findings that turtle
photoreceptors neither synthesize nor take up GABA (16). The concentrations of all other amino acids measured in isolated retinal cells and in
photoreceptor-rich fractions were similar within 50%. In particular,
the concentrations of free aspartic and glutamic acids in isolated photoreceptor fractions were not higher than those found in the retina. As
in other vertebrate retinas, the concentration of taurine was much
higher than those of other amino acids in both the photoreceptors and in
the retina (1,5,14,25,34).

TABLE 3 CONTENT OF FREE AMINO ACIDS IN PHOTORECEPTOR-RICH FRACTIONS AND IN
ISOLATED CELLS OF THE TURTLE RETINA

	Photoreceptor-rich fractions	Isolated retinal cells
Glutamic acid	18.3 ± 0.18	26.9 ± 0.26
Aspartic acid	5.1 ± 0.35	8.9 ± 0.84
Glycine	11.2 ± 1.5	12.9 ± 0.14
γ-Aminobutyric acid	<0.3	14.8 ± 0.3
Taurine	126 ± 7.5	164 ± 5.9
Threonine	2.41 ± 0.16	2.17 ± 0.11
Serine	8.5 ± 0.20	6.8 ± 0.71
Glutamine	16.1 ± 0.81	28.9 ± 1.4
Alanine	7.6 ± 1.0	5.4 ± 0.04
Valine	2.6 ± 0.22	1.4 ± 0.05
Isoleucine	1.8 ± 0.04	1.0 ± 0.01
Leucine	2.6 ± 0.6	1.9 ± 0.03
Tyrosine	10.1 ± 0.86	9.6 ± 0.12
Phenylalanine	2.9 ± 0.22	0.9 ± 0.02
Histidine	1.8 ± 0.13	2.8 ± 0.06
Lysine	3.3 ± 0.29	1.3 ± 0.03
Arginine	3.9 ± 0.27	1.5 ± 0.09

Values are means ± s.e.s in μmol/g protein for 5 separate preparations.

Acetylcholine

Acetylcholine (ACh) and its metabolic enzymes, choline acetyltransferase and acetylcholinesterase, have been demonstrated in retinas of

many different species (11). The sites of the cholinergic synapses, however, have not been identified unequivocally. Because cholinergic neurons usually contain high levels of ACh (7,23,28,29), we have measured the concentration of ACh in isolated photoreceptors and in the retina.

The endogenous levels of ACh in isolated photoreceptors, in retinal cells before and after sedimentation, and in the retina were determined by the method of McCaman and Stetzler (24) for measuring sub-picomole quantities of ACh. Retinas or isolated cells were homogenized in formic acid and acetone (15:85 by vol), extracted with sodium tetraphenol boron in 3-heptanone (5mg/ml) and then 0.4 M HCl. The extracts were dried and first incubated for 15 min with ATP and choline kinase (ATP:choline phosphotransferase EC 2.7.132) to covert all free choline to phosphorylcholine. Acetylcholinesterase (AChE, EC 3.1.1.7) and $(\gamma-^{32}P)$ ATP (New England Nuclear Corp., Boston, MA) were then added to the mixture and incubated for another 15 min to hydrolyze ACh and phosphorylate the resulting choline. The ^{32}P-phosphorylcholine was separated on an ion-exchange column under alkaline conditions and the radioactivity was measured in a Beckman liquid scintillation counter

TABLE 4 ACh, Ch, ChAT AND AChE IN ISOLATED PHOTORECEPTORS

	Photoreceptors	Photoreceptor-rich fraction	Sedimented retinal cells	Retina
Acetylcholine*	1.96 ± 0.36	1.12 ± 0.34	0.74 ± 0.05	1.05 ± 0.10
Choline*	—	0.36 ± 0.12	—	0.44 ± 0.08
Choline acetyltransferase†	4.0 ± 0.70‡	—	18.3 ± 2.6§	15.9 ± 2.40§
Acetylcholinesterase†	—	2773 ± 209§	7022 ± 795§	8549 ± 641§

* Values are means ± s.d.s of $\mu mol/g$ protein for 3–4 experiments; † Values are means ± s.d.s of μmol product/h/g protein for 3–4 experiments. ‡ This value is from LAM (1975). § These values are from SARTHY and LAM (1978).

Using this procedure, the ACh contents in turtle retinas, isolated retinal cells, photoreceptor-rich fraction and identified photoreceptors were measured. Our results show that turtle photoreceptors have an ACh concentration similar to that of the whole retina (Table 4). In addition, the contents of choline in photoreceptors and retinas are similar, whereas cell free extracts of photoreceptors contain about 25% of the specific activities for choline acetyltransferase (ChAT) and acetylcholine esterase (AChE) found in the retina.

α-Bungarotoxin Binding Sites in Goldfish and Turtle Retinas

α-Bungarotoxin (α-BTX) has been shown to be a specific ligand for certain nicotinic receptors (3). Using ^{125}I-labeled α-BTX, Yazulla and Schmidt (36) showed by light microscope autoradiography that (^{125}I)α-BTX binding sites are associated with both the outer and inner plexiform layers of goldfish and turtle retinas. By means of electron microscope autoradiography, Schwartz and Bok (31) demonstrated that within the outer plexiform layer of the goldfish retina (^{125}I)α-BTX binding sites are most likely situated on small bipolar cell dendritic processes that invaginate rod and cone synaptic terminals, and on large bipolar cell dendritic processes more proximally situated in the outer plexiform layer (Fig. 2). Horizontal cell processes did not appear to be labeled. Additionally, they showed that the (^{125}I)α-BTX binding was suppressed by 10^{-5} M nicotine or 10^{-5} M unlabeled α-BTX.

More recently, Schwartz and Bok (private communications) have extended

FIG. 2. Electron micrograph of the distal region of a goldfish retina incubated with (^{125}I)α-bungarotoxin. As described in detail by Schwartz and Bok (31), the label is most probably situated on certain bipolar dendritic processes (arrow), but not on horizontal cell processes. Scale: 1 μm. Micrograph from D. Bok, UCLA. (See p. 72.)

FIG. 3. Electron micrograph of the distal region of a turtle retina incubated with (^{125}I)α-bungarotoxin. In this retina, the label is also most probably situated on certain bipolar dendritic processes (arrows) but not on horizontal cell processes. Scale: 1 μm. Unpublished micrograph from Bok. (See p. 72.)

---

their analyses to the turtle retina. Their results strongly suggest that similar to the goldfish retina, specific (^{125}I)α-BTX binding sites are probably associated with membrane receptors of certain bipolar dendritic processes in the outer plexiform layer (Fig. 3). Clearly identified processes of horizontal cells are free of label.

Glutamate and Aspartate Uptake and Release in Human and Other Primate Retinas

Unlike cold-blooded vertebrates, the evidence for cholinergic synapses in the outer plexiform layer is considerably weaker in mammalian retinas. Because aspartic and glutamic acid are the leading transmitter candidates for mammalian photoreceptors, we have attempted to determine whether these cells possess specific, high-affinity uptake systems for these amino acids. We chose primate (Rhesus macaque and human) retinas for these studies because these retinas have rods and three types of cones; thus, all four types of photoreceptors can be studied simultaneously. Isolated retinas were incubated for 15 min in oxygenated mammalian Ringer's solution supplemented with 10% fetal calf serum and 1 to 10 μM L-(^{3}H)aspartic acid (sp. act. 17.3 Ci/mmole) or L-(^{3}H)glutamic acid (sp. act. 43 Ci/mmole). The retinas were then processed for light and electron microscope autoradiography as described elsewhere (17). As shown in Fig. 4, under our experimental conditions, there is no obvious selective uptake of glutamate into any population of cells in the human retina. Specifically, all the photoreceptors as well as many other retinal cells are labeled. However, when human retinas were incubated with (^{3}H)aspartic acid under the same conditions, glial (Müller) cells and a subpopulation of photoreceptors are selectively labeled (Figs. 5 and 6). Specifically, all the rods but none of the cones appear to be labeled. This finding is illustrated more vividly in over-exposed autoradiographic sections of aspartate and glutamate uptake (Figs. 6 and 7). As shown in Fig. 6, human retinas incubated with (^{3}H)aspartate show

---

FIGS. 4-7. Autoradiographic sections of isolated human retinas incubated for 20 min with 200 μl Ringer's solutions each containing the following (^{3}H) neurotransmitter candidates. 4: (^{3}H)glutamate--label is most heavily associated with all photoreceptors and Müller cells. 5: (^{3}H)aspartate--label is most heavily associated with Müller cells and rod photoreceptors. None of the cone photoreceptors (arrows) are labeled. 6: (^{3}H)aspartate--overexposed section showing more clearly the differential labeling between rods and cones (arrows). 7: (^{3}H)glutamate--overexposed section showing that both rods and cones are heavily labeled. Scales = 30 μm. (18). (See p. 73.)

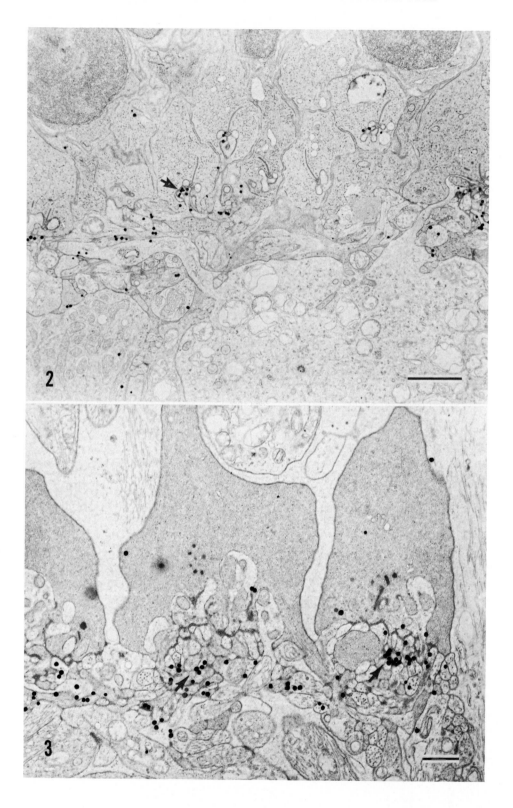

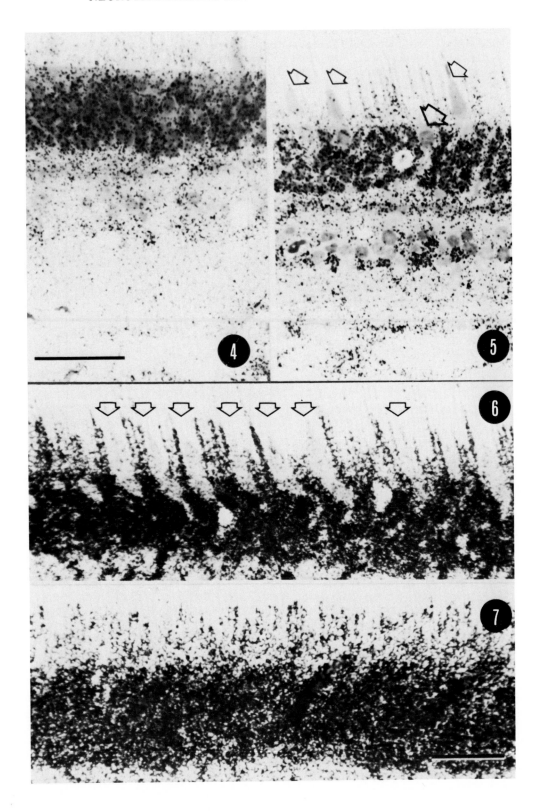

selective labeling over rods but not over cones. In contrast, both rods and cones are labeled to similar extents in human retinas incubated with (^{3}H)glutamate (Fig. 7). We have also shown that cones but not rods of <u>macaque</u> retinas selectively take up (^{3}H)aspartate, and that all photoreceptors in this retina take up (^{3}H)glutamate (D.M.K. Lam, unpublished data). We must, however, emphasize that we have not yet analyzed the chemical identifications of the labeled substances in the photoreceptors and, therefore, do not know if the label in these cells is still predo-

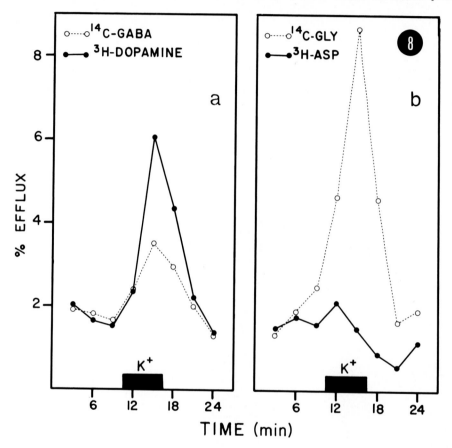

FIG. 8. a: Simultaneous efflux of (^{14}C)GABA and (^{3}H)dopamine in response to membrane depolarization induced by isotonic K^+-rich medium (56 mM K^+). Retinas were incubated for 20 minutes with a medium containing (^{14}C)GABA and (^{3}H)dopamine, washed with unlabeled medium for 1 h prior to incubation with the K^+-rich medium. The medium contained 2 mM GABA and β-alanine to inhibit GABA uptake. Under these conditions, both (^{14}C)GABA and (^{3}H)dopamine were released in Ca^{++}-dependent manner. b: Simultaneous efflux of (^{14}C)glycine and (^{3}H)aspartic acid in response to membrane depolarization induced by isotonic K^+-rich medium (56 mM K^+). The medium contained 0.1 mM D-aspartate to inhibit aspartate uptake. Under conditions similar to those described in Figure 10, (^{14}C)glycine is readily released but there is no significant efflux of (^{3}H)aspartate in response to K^+ stimulation. (18).

minantly aspartate. Nonetheless, these results demonstrate a chemical difference between rods and cones with respect to their abilities in accumulating or metabolizing aspartate.

We next examined the ability of human rod cells to release the accumulated (^{3}H)aspartate in response to increased K$^+$-concentration in the medium. In retinas incubated with (^{3}H)aspartate and (^{14}C)glycine simultaneously, about 16% of the (^{14}C)glycine taken up by the retina is released by two 3-minute incubations in K$^+$ rich Ringer's solution (Fig. 8). This release is blocked by the presence of 5 mM Co^{++} in the medium. However, very little (^{3}H)aspartate is released under the identical conditions. Moreover, even the addition of 0.1 mM L-aspartate or D-aspartate to inhibit aspartate uptake fails to elicit K$^+$-stimulated efflux of (^{3}H)aspartate from the retina. Further studies are therefore required to examine whether depolarization of the retina leads to release of the aspartate taken up by the rod photoreceptors.

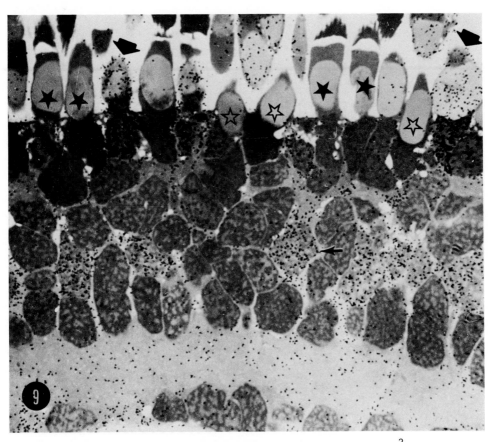

FIG. 9. Autoradiographic section of the accumulation of L-(^{3}H)aspartate by cells in the <u>Necturus</u> retina. Müller cells (horizontal arrow) are labeled. A subpopulation of presumed cones is heavily labeled (oblique arrows). Rods (closed stars) and many presumed cones (open stars) are, however, not labeled. Original.

FIG. 10. Comparisons of L-(^{3}H)aspartate, L-(^{3}H)glutamate and D-(^{3}H)aspartate accumulation by cells in the goldfish retina. a: Glutamate uptake by rods in a quartered retina. Rod ellipsoids (circles) and rod nuclei (arrowheads) are labeled. A distinct boundary beyond which rod nuclear labeling declines dramatically is indicated by a row of asterisks. The retina was dark-adapted to retain as many rods as possible, so that most cones have extended beyond the rod tips. A remaining blue-sensitive cone (B) shows no uptake. 100 µCi/ml L-(^{3}H) glutamate, 5 µM; 1/2 µm section; exposed 14 days to reveal sufficient grains without obscuring cell somas; understained with Toluidine Blue so as not to obscure grains over the rods. b: L-(^{3}H)glutamate localization in a 500 µm wide, light-adapted retinal slice. Cones (R,G) show a small amount of uptake but rod nuclei (arrowheads) in the outer nuclear layer (ONL) are well-labeled. Cone synaptic terminals (squares) are not labeled despite the fact that the overall label density in the outer plexiform layer (OPL) has increased over that seen in larger retinal pieces. The remainder of the retina is diffusely labeled with grain accumulations over bipolar cell nuclei in the inner nuclear layer (INL) and over the inner plexiform layer (IPL) in general. PE, pigment epithelium. 100 µCi/ml L-(^{3}H)glutamate, 5 µM; 1/2 µm section; exposed 8 days; stained with Toluidine Blue to reveal cone terminals and proximal retinal cells. This staining procedure overstains rods so that ellipsoid-specific labeling is difficult to observe. c: Higher magnification view of a 1/2 µm section immediately adjacent to that shown in (b). Here the localization of label to rod nuclei (arrowheads) is clear, whereas the labeling of cone nuclei (CN) and their surrounding cytoplasm is low to moderate. The distal-to-proximal label gradient observed in rod nuclei shown in (a) is dramatically decreased in the slice preparation but not abolished. Cone terminals (squares) show no labeling whereas clusters of rod terminals (circles) between cone endings are heavily labeled. The spaces between rod nuclei, occupied primarily by Müller's cell processes, are free of label. d: L-(^{3}H)aspartate accumulation by cones. Cones identifiable as red-sensitive (R) or green-sensitive (G) show light but distinct localization of L-(^{3}H)aspartate. Blue-sensitive cones (B) and rods (circles) do not. No selective labeling of cone terminals (squares) is observed. Neural elements in the remainder of the retina do not accumulate L-(^{3}H)aspartate. 100 µCi/ml L-(^{3}H)aspartate, 15 µM; 1/2 µm section; exposed 14 days for direct comparison with (a). e: D-(^{3}H)aspartate uptake by cones and glial cells. This preparation was incubated three times as long as all other experiments (30 min versus 10 min) to assess how selectively red-sensitive and green-sensitive (RG) cones could be labeled. Despite the long incubation, rods and blue-sensitive cones (B) have six-fold fewer grains than red-sensitive and green-sensitive cones. Also heavily labeled are Müller's cells (M), which showed little uptake of L-(^{3}H)glutamate. N, cone nucleus. 100 µCi/ml D-(^{3}H)aspartate, 10 µM; 1/2µm section; 24 day exposure. f: L-(^{3}H)aspartate accumulation by Müller's cells (M) and exclusion by horizontal cells (H), bipolar cells (B), amacrine cells (A), ganglion cells (GC) and bipolar cell axon terminals (BAT). 1 mCi/ml L-(^{3}H)aspartate 150 µM; 1/2 µm section; 20 day exposure. From Marc and Lam (22).

Aspartate Uptake by Necturus Retina

We have also studied the uptake of (^{3}H)aspartate in the Necturus

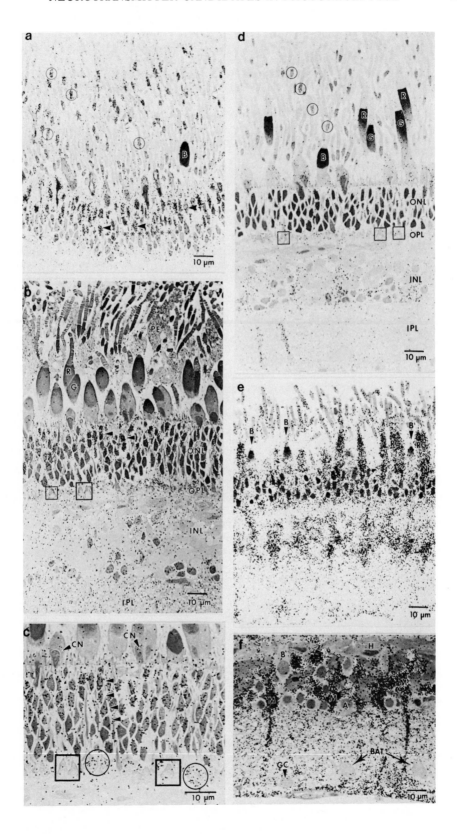

retina. Under our experimental conditions, a small number of photoreceptors take up (^{3}H)aspartic acid. Unlike the human retina, the rods in
this retina do not accumulate (^{3}H)aspartic acid and only a subpopulation
of the cones take up this compound (Fig. 9). This uptake pattern is,
therefore, different from that found in primate photoreceptors.

Aspartate and Glutamate Uptake by Goldfish Retina

We have also examined the uptake of acidic amino acids by photoreceptors in the goldfish retina using light microscopic autoradiography
(22, and Fig. 10). Isolated retinas were incubated in media containing
micromolar amounts of L-(^{3}H)aspartate or L-(^{3}H)glutamate. We found that
rods accumulate L-(^{3}H)glutamate with a high-affinity transport system;
they exhibit a glutamate:aspartate selectivity ratio of 30:1. When incubated in 1-10 μM L-(^{3}H)glutamate, rods label more densely than cones.
A unit area of rod membrane transports glutamate 30 times better than a
unit area of cone membrane. Red-sensitive and green-sensitive cones show
accumulation of L-(^{3}H)aspartate, D-(^{3}H)aspartate and L-(^{3}H) glutamate,
apparently with high affinity, but with little selectivity. Because rods
have poor aspartate uptake, red-sensitive and green-sensitive cones may
be preferentially labeled with L-(^{3}H)aspartate or D-(^{3}H)aspartate. Blue-
sensitive cones show no uptake of L-(^{3}H)aspartate, D-(^{3}H)aspartate or
L-(^{3}H)glutamate other than that attributable to low affinity transport.

Taken together, our results indicate that both rods and cones possess
somatic high-affinity transport sites and that various cell types in the
goldfish retina can clearly discriminate between glutamate and aspartate,
unlike acidic amino acid transport systems described in mammalian brain
(6). The possibility that these amino acids or their analogues may be
photoreceptor transmitters must, however, await further studies.

CONCLUSION

Our results, considered with those of other investigators, point to
acetylcholine as a likely transmitter candidate for cephalopod photoreceptors. The transmitters used by vertebrate photoreceptors are, however, very much less certain. Our findings indicate strongly that even
within the same species, different types of photoreceptors may use different neurotransmitters. In addition, the transmitter(s) used by
mammalian rods may be different from the transmitter(s) used by rods of
lower vertebrates, such as teleosts and amphibians. To date, not only
are the vertebrate photoreceptor transmitters still unknown, but the
physiological significance of deploying different transmitters for different photoreceptors in the same species as well as throughout evolution
is unclear.

ACKNOWLEDGMENTS

We thank Professors D. Bok and I.R. Schwartz who provided the micrographs for figures 2 and 3. Tables 1 and 2 are from reference 19;
tables 3 and 4 are from reference 30. We thank Gayle H. Shipp for
typing the manuscript. J.M.F. is a National Research Service Awardee,
D.M.K.L. and J.G.H. are recipients of Research Career Development Awards
of the National Eye Institute. D.M.K.L. is an Olga K. Weiss Scholar of
Research to Prevent Blindness, N.Y. This work was supported by grants
EY05508 (J.M.F.), EY02423 (D.M.K.L.), and EY02363 (J.G.H.) from the

National Eye Institute and from the Retina Research Foundation (Houston).

REFERENCES

1. Berger,S.J.,McDaniel,M.L.,Carter,J.G., and Lowry,O.H.(1977): J. Neurochem., 28:159-163.

2. Cervetto,L., and MacNichol,E.F.,Jr.(1972): Science, 178:767-768.

3. Chang,C.C., and Lee,C.Y.(1963): Arch. Int. Pharmacodyn. Ther., 144:241-257.

4. Cohen,A.(1972): J. Comp. Neurol., 147:379-398.

5. Cohen,A.J.,McDaniel,J., and Orr,H.(1973): Invest. Ophthalmol., 12: 686-693.

6. Davies,L.P., and Johnston,G.A.R.(1976): J. Neurochem., 26:1007-1014.

7. Dowdall,M.J.,Fox,G.,Wachtler,K.,Whittaker,V.P., and Zimmerman,H. (1976): Cold Spring Harbor Symp. Quant. Biol., 40:65-81.

8. Dowling,J.E., and Ripps,H.(1972): J. Gen. Physiol., 60:698-719.

9. Drukker,J., and Schadé,J.P.(1964): Neth. J. Sea Res., 2:155-182.

10. Gerschenfeld,H.M., and Piccolino,M.(1977): Nature (London), 268:257-259.

11. Graham,L.T.Jr.(1974): In: The Eye:Comparative Physiology, Vol. 6, edited by H. Davson and L.T. Graham,Jr. pp. 283-342, Academic Press, New York.

12. Gray,E.G.(1970): J. Cell Sci., 7:203-215.

13. Hildebrand,J.G.,Barker,D.,Herbert,E., and Kravitz,E.(1971): J. Neurobiol., 2:231-246.

14. Kennedy,A.J., and Voaden,M.J.(1974): J. Neurochem., 23:1093-1095.

15. Lam,D.M.K.(1972): Proc. Natl. Acad. Sci. USA, 69:1987-1991.

16. Lam,D.M.K.(1976): Cold Spring Harbor Symp. Quant. Biol., 40:571-579.

17. Lam,D.M.K., and Hollyfield,J.G.(1980): Exp. Eye Res., 31:729-732.

18. Lam,D.M.K.,Frederick,J.M., and Hollyfield,J.G.(1981): In: The Structure of the Eye, edited by Joe G. Hollyfield, Elsevier/North Holland, Inc., New York (in press).

19. Lam,D.M.K.,Wiesel,T.N.,Kaneka,A.(1974): Brain Res., 82:365-368.

20. Lund,R.D.(1966): Exp. Neurol., 15:100-112.

21. Mains,R., and Patterson,P.(1973): J. Cell Biol., 59:329-345.

22. Marc,R.E., and Lam,D.M.K.(1981): Proc. Natl. Acad. Sci. USA, in press.

23. McCaman,R.E., and McCaman,M.W.(1976): In: Biology of Cholinergic Function, edited by A.M. Goldberg and I. Hanin, pp. 485-513. Raven Press, New York.

24. McCaman,R.E., and Stetzler,J.(1977): J. Neurochem., 28:669-671.

25. Morales,H.P.,Klethi,J.,Ledig,M., and Mandel,P.(1972): Brain Res. 41:494-497.

26. Murakami,M.,Ohtsu,K., and Ohtsuka,T.(1972): J. Physiol. (Lond), 227:899-913.

27. Murakami,M.,Ohtsuka,T., and Schimazaki,H.(1975): Vision Res., 15: 456-458.

28. Saelens,J.K., and Simke,J.P.(1976): In: Biology of Cholinergic Function, edited by A.M. Goldberg and I. Hanin, pp. 661-681. Raven Press, New York.

29. Sanes,J.R., and Hildebrand,J.G.(1976): Dev. Biol., 52:105-120.

30. Sarthy,P.V., and Lam,D.M.K.(1979): J. Neurochem., 32:455-461.

31. Schwartz,I.R., and Bok,D.(1979): J. Neurocytol., 8:53-66.

32. Sugawara,K., and Negishi,K.(1973): Vision Res., 13:2479-2489.

33. Tonosaki,A.(1965): Z. Zellforsch., 67:521-532.

34. Voaden,M.J.,Lake,N.,Marshall,J., and Morjaria,B.(1977): Exp. Eye Res., 25:249-257.

35. Wu,S.M., and Dowling,J.E.(1978): Proc. Natl. Acad. Sci. USA, 75: 5205-5209.

36. Yazulla,S., and Schmidt,J.(1976): Vision Res., 16:878-880.

37. Young,J.Z.(1962): Philos. Trans. R. Soc. Lond. (Biol), 245:1-58.

Visual Cells in Evolution, edited by Jane A. Westfall,
Raven Press, New York © 1982.

Evolution of Synapses in Visual Cells

Jane A. Westfall

Department of Anatomy and Physiology, Kansas State University, Manhattan, Kansas 66506

As a general rule structure correlates with function and both evolve
over time. Based on this premise one can formulate logical hypotheses
for evolution of various organelles within different animal phyla. The
purpose of this review is to compare ultrastructural features of visual
cell synapses in representatives of four diverse phylogenetic groups,
emphasizing (1) the components of the synaptic complex, and (2) the pattern of synaptic connectivity between photoreceptors and their second
order neurons. Because the literature is poor on invertebrate photoreceptor synapses other than those of arthropods, I have selected examples from only three groups of invertebrates (coelenterates, molluscs
and arthropods) to compare with vertebrate visual cell synapses.

COELENTERATES

Coelenterates, or more specifically those animals assigned to the
Phylum Cnidaria, belong to the phylogenetically most primitive group of
animals to possess a nervous system. Some coelenterates, such as certain hydromedusae and scyphomedusae, have ocelli where tentacles contact
the margin of the bell. Such marginal sense organs may be either simple
open epithelial pits or closed structures containing a lens and layered
retina. In their simplest form ocelli are composed of electron lucent
receptor cells interspersed among dense pigmented cells and surrounded
by large vacuolate peripheral neurons (27). Only three ultrastructural
studies have been done on the synaptic connections in coelenterate visual cells (27, 32, 33). In general, chemical synapses are characterized
by the presence of presynaptic vesicles and a pair of uniformly apposed
membranes with pre-and post-synaptic specializations. Receptor cells in
the hydromedusan Spirocodon saltatrix have somatic and axonal synapses
with second order neurons that in turn form less conspicuous contacts
back onto the original receptor cells (27). In the cubomedusan Tamoya
busaria processes of second order neurons are invaginated into tapered
ends of sensory cells forming putative synaptic contacts with closely
associated 80 nm diameter vesicles (33).

Paired electron dense membranes separated by a 20-30 nm cleft with
vesicles on one side of the contact are characteristic of visual cell
synapses in coelenterates as well as other animals. In Spirocodon the
pre-and post-synaptic paramembranous densities are of approximately
equal thickness so that the synaptic membranes appear symmetrical with

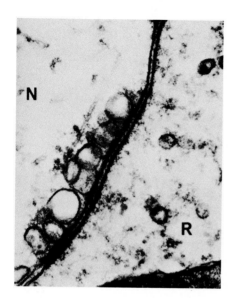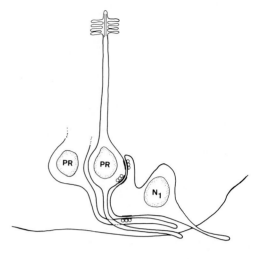

FIGS. 1 and 2. 1: Electron micrograph of feedback synapse between second order neuron (N) and photoreceptor cell (R) in a coelenterate, Spirocodon saltatrix, with paramembranous densities, cross filaments in
cleft, and a row of large membrane-associated vesicles. X 65,000. From
Toh et al. (27). 2: Diagrammatic representation of synaptic contacts
between photoreceptor cells (PR) and second order neuron (N_1) in coelenterate ocellus illustrating somatic and axonal contacts of receptor
cells onto the second order neuron and its feedback contact with a receptor cell. Original.

fine cross filaments filling the cleft. Some of the vesicles are in contact with dense projections on the presynaptic membrane (Fig. 1). Receptor cells have larger numbers of smaller-sized vesicles (80-110 nm
in diameter) than the second order neurons in which only a few vesicles
(up to 200 nm in diameter) are associated with presynaptic dense projections. Synaptic contacts of both receptor cells and second order
neurons may be either somatic or axonal in location (27).

Based on a limited amount of published data, I postulate that the
basic pattern of synaptic connectivity of coelenterate visual cells is
one in which photoreceptor cells have somatic and axonal contacts onto
second order neurons that in turn have feedback contacts on the photoreceptor cells (Fig. 2). To date, no electrical synapses, i.e., gap
junctions, have been described between visual cells in coelenterates,
although I believe future investigations will demonstrate their presence.

MOLLUSCS

Organs of vision in the Phylum Mollusca range from simple ocelli a-
round the mantles of certain clams such as Tridacna to the complex image
forming eyes of squid. Very little ultrastructural study has been done
on the synaptic connections of visual cells in molluscs; hence, I shall
confine my observations to one study on photoreceptor cell synapses in

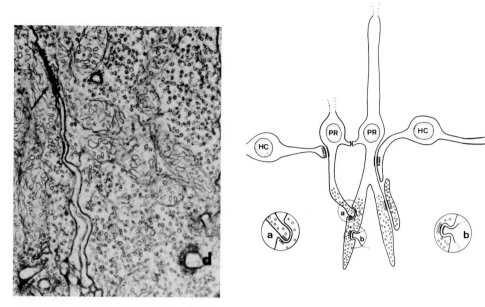

FIGS. 3 and 4. 3: Photoreceptor terminal in a mollusc, Loligo pealei, with numerous synaptic vesicles, an invaginated synapse on dendrite (d) of a second order neuron, and an interaxonal receptor cell contact (arrow). X 18,000. From Cohen (5). 4: Diagrammatic representation of pattern of synaptic connectivity of photoreceptor cells (PR) and horizontal cells (HC) in squid eye illustrating invaginated axo-axonal (a) and axodendritic (b) contacts of receptor cells, en passant and terminal synaptic contacts of horizontal cells, gap junctions between somatic processes, and modified chemical type synaptic contacts between axonal processes of photoreceptor cells. Original.

the common squid, Loligo pealei.

The rhabdomeric photoreceptor cells have long saclike terminals abutting the outer plexiform layer of nerve fibers (5). Within the vesicle-filled terminals are rows of invaginated synapses. These appear as axo-dendritic-type contacts on processes of second order neurons (Figs. 3,4). The dendritic invaginations are small; they are characterized by a horseshoe-shaped crest of 40 nm diameter vesicles in close contact with the presynaptic density of the synaptic membrane complex. Paramembranous densities and a cleft width typical of chemical synapses are present.

Other types of putative synaptic contacts also occur in the rhabdomeric terminals. One specialized contact between axon terminals is characterized by vesicles on one side of the junction and a flattened cisterna on the other side (arrow, Fig. 3). The apposed membranes are separated by a 20-30 nm cleft traversed by fine filaments typical of chemical synapses. Another axo-axonal contact has interdigitated processes with paired elongate submembranous cisternae on either side of a 2.5-4 nm wide intercellular gap (a, Fig. 4). Synaptic vesicles cannot adhere to the paired membranes owing to the close apposition of the narrow cisternae on both sides of an otherwise typical gap junction. In addition to these unusual electrical junctions, typical gap junctions are present between processes from perikaryal regions of receptor cells

(5).

The second order neurons in the squid eye have horizontal processes with en passant synaptic contacts characterized by clusters of small clear vesicles at a synaptic membrane complex. Such contacts are characteristic of feedback or efferent synapses of second order neurons on the rhabdomeric photoreceptors. There are similar axosomatic contacts onto rhabdomeric photoreceptors in the octopus retina where clusters of agranular synaptic vesicles (30-50 nm in diameter) are aggregated near a 28 nm wide synaptic cleft with dense material (10).

Based mainly on a study of the squid eye, I postulate that the principal pattern of connectivity of mollusc photoreceptors is one in which photoreceptors have multiple axodendritic synapses on second order neurons that in turn have feedback efferent contacts on the receptor cells (Fig. 4). Squid visual cells have axo-axonal junctions with either submembranous cisternae and vesicles or a pair of cisternae but no membrane-associated vesicles. In addition to those specialized junctions there are typical gap junctions between cytoplasmic extensions of receptor cells. Both the patterns of connectivity and the types of synaptic contacts are more complex in squid visual cells than in coelenterate photoreceptors.

ARTHROPODS

The Phylum Arthropoda has a greater number and diversity of species than any other phylum in the animal kingdom. The majority of species occur in the Class Insecta where both simple ocelli and compound eyes are present. The compound eye is made up of ommatidia composed of a specific number of light sensitive cells in each facet. In the common house fly, Musca domestica, there are eight light-sensitive cells (retinula cells) in each ommatidium and a total of 44,800 visual cells in the paired compound eyes (24). The retinula cells form short or long axons terminating on second order neurons in the lamina (first neuropile) or in the medulla of each eye. The synaptic connections within the laminar region of the fly eye have been studied by numerous investigators (2,3,20,25,29). They are characterized by T-shaped or table type presynaptic densities (28) and are associated with one or more second order neurons in the lamina (L1 and L2, Fig. 5). A serial section reconstruction of the receptor cell-laminar cell synaptic complex demonstrates a presynaptic bar (35 nm x 220-260 nm) lying parallel to the synaptic membrane and separated from it by a 10 nm wide space (4). The bar is covered by a 10 nm thick presynaptic plate (giving it a table shape) from which it is separated by about 10-15 nm. The synaptic vesicles (30-40 nm in diameter) are regularly arranged along the long axis of the presynaptic bar. The uniform 18 nm wide synaptic cleft may involve four postsynaptic cells, two of which contain postsynaptic bags and whiskers (see Fig. 4a in 4). The feedback synapses of laminar cells on retinula cells have similar T-shaped contacts and divergent dyad or triad associations (Figs. 5,6).

Synapses of retinula cells in the dragonfly, Sympetrum rubicundulum, contain elongate presynaptic specializations and have either divergent dyad or triad arrangements of postsynaptic elements (1). The presynaptic density of the dragonfly retinula cell is separated from the synaptic membrane by a 10 nm space. The density itself can be resolved into two subelements that follow the curvature of the presynaptic ridge. There are synaptic vesicles specifically clustered along the density and uni-

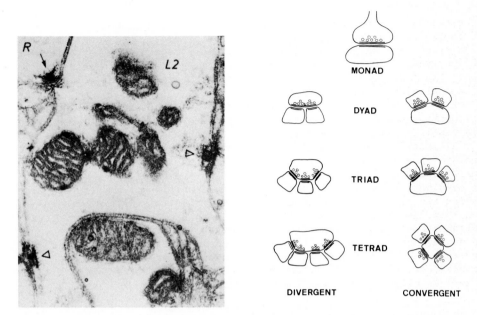

FIGS. 5 and 6. 5: Specialized presynaptic densities with T-shaped or table configuration at both a photoreceptor cell synapse (dark arrow) and at laminar cell synapses (clear arrowheads) in an arthropod, Musca domestica. X 45,000. From Burkhardt and Braitenberg (4). 6: Diagrammatic representation of synaptic connections in arthropod visual cells illustrating divergent and convergent patterning of dyad, triad and tetrad configurations. Original.

formly distributed throughout the terminal. Intracleft densities and postsynaptic specializations occur in all elements of the triad. Retinula cells may be presynaptic to monopolar neurons MI and MII in triadic arrangements whereby MII forms the central postsynaptic process and MI forms the lateral processes. Such a pattern of central and lateral postsynaptic processes from different second order neurons is also found in the vertebrate eye. In the dragonfly, however, the central monopolar neuron may form feedback synapses upon a pair of retinular terminals, the pattern of connection being dependent on the stratification of terminal depths and the specificity of retinular connections.

Divergent triad synapses are also found in the compound eye of the crayfish Procambarus where a central postsynaptic element is flanked by a pair of lateral processes from different neurons (11). The receptor cell is characterized by a bar-shaped presynaptic density similar to that of the dragonfly. The terminal expansion contains clear vesicles (40-50 nm in diameter) that are also aligned along the presynaptic density. Similar presynaptic bars have been described in the lobster (12) and other arthropods. The variations among species are more quantitative than qualitative in nature and need to be reviewed at a later date.

In addition to the synaptic bar structures characteristic of arthropod visual cells there are conventional chemical synapses and gap junctions. Spider eyes have conventional synapses between photoreceptor axons and the rhabdomere-bearing part of a photoreceptor cell (17). Gap junctions have been demonstrated between short retinula cell axons in the lamina of the fly (20); they are characterized by a 2-4 nm intercellular gap and an accumulation of semi-dense material on both sides of the junctions. The plaques are about 0.3 μm in diameter in thin section. Gap junctions have also been reported between visual cells and between glial and visual cells in Limulus (16).

Arthropods have a complexity of presynaptic structure and connectivity not found in other invertebrate photoreceptors. The form of the presynaptic density can vary in different arthropod species but the general pattern appears to be one of a specialized presynaptic component separate from the synaptic membrane. The long axis of the presynaptic density has a characteristic row of vesicles and parallels the presynaptic membrane. Postsynaptic membranes have densities or structures paralleling the presynaptic components. In addition to simple or conventional (monad) synaptic specializations there are dyad, triad and tetrad configurations where one presynaptic element may impinge on more than one postsynaptic element (divergent) or convergent patterns in which more than one presynaptic element may converge on a single postsynaptic element (Fig. 6). Gap junctions have been reported between visual cells as well as between glial and visual cells.

VERTEBRATES

The structures and functions of many vertebrate eyes have been studied, making interspecific comparisons possible (21,23). A typical vertebrate eye contains rods and cones that form synaptic contacts with two types of second order neurons: bipolar cells and horizontal cells. Bipolar cell processes make both invaginated and superficial contacts onto receptor terminals. Horizontal cells have dendritic and axonal contacts with receptor terminals. A receptor cell terminal contains numerous clear synaptic vesicles (40-50 nm in diameter) and a presynaptic ribbon or lamella surrounded by a halo of vesicles at the putative active zone (Fig. 7). The elongate rodlike lamellar structure has a dense core and is arranged perpendicular to the presynaptic membrane from which it is slightly separated by the arciform lamella (15). Developmental and freeze-fracture studies of visual cell synapses indicate their ultrastructural similarity to other chemical synaptic complexes (18,22). The receptor terminals (rod spherules and cone pedicles) have multiple parallel invaginations that envelop the terminal processes of both bipolar and horizontal cells. According to Shepherd (21: p. 193), "In primates, the invaginations in the rod spherule contain a central process of the dendrite of a rod bipolar cell, flanked by two processes from the 'axonal' arborizations of horizontal cells...Cone pedicles have invaginations in which a dendrite from an invaginating midget bipolar cell is flanked by two processes from the 'dendritic' arborization of horizontal cells...In the central retina, a single invaginating midget bipolar cell connects to a single cone pedicle by means of 10-25 triads...Flat midget bipolar cells connect to approximately 6 cones, whereas rod bipolar cells connect to 10-50 rod spherules."

Obviously such complexity makes the pattern of connectivity in vertebrate visual cells difficult to trace. Although parts of the pre- and post-synaptic membranes show specialized thickenings, it is unclear

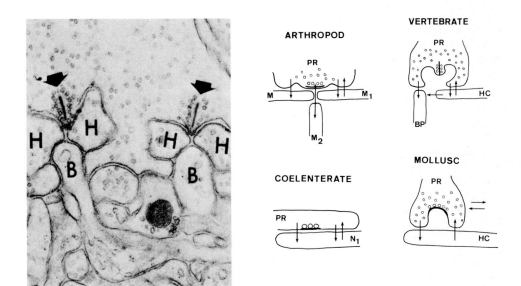

FIGS. 7 and 8. 7: Parallel invaginated ribbon synapses (arrows) in cone cell of rhesus monkey with paired horizontal cell processes (H) lateral to single central bipolar cell process (B). X 42,000. From Dowling (7). 8: Summary diagram illustrating evolutionary similarities and differences in synaptic structure and connectivity patterns of visual cells. Arrows indicate structural polarity of connectivity patterns between photoreceptors (PR) and a second order neuron (N_1), monopolar neurons (M, M_1, M_2), horizontal cells (HC) and a bipolar cell (BP) in examples from four phylogenetically distinct groups of animals. Original.

whether the bipolar cell process is postsynaptic to the photoreceptor, the horizontal cell processes, or both (7,23). Superficial contacts from bipolar processes onto receptor terminals may show membrane specializations without associated vesicles (7). Horizontal cell processes have conventional synapses but they are not associated with rod sperules or cone pedicles (8). In some vertebrate retinas however bidirectional interactions between receptors and horizontal cells have been indicated (23).

Gap junctions have been reported between vertebrate cone pedicles and between cone pedicles and rod spherules (14). In the axolotl retina gap junctions with a 2-4 nm intercellular gap are present between radial fins of neighboring rods and also between neighboring rod and cone cells (6). Freeze-fracture preparations of monkey retina revealed gap junctions between cone pedicles (19). The gap junctions are characterized by aggregations of intramembranous particles approximately 8-14 nm in diameter, present in the A face of the fracture membrane.

Presynaptic ribbons, characteristic of vertebrate visual cell synapses, are also found in mammalian pineal organs where other photoreceptive structures have been lost in the process of evolution. They may occur as a perpendicular rod with a halo of vesicles at the pinealocyte membrane or lie free in the cytoplasm where they have been shown to vary in number with a circadian rhythm (13,26,30).

In general, vertebrate rod and cone terminals are characterized by the presence of ribbon synapses. The ribbons or lamellae are oriented perpendicular to the presynaptic membrane and have a "halo" of vesicles. There is also increased complexity of the patterns of synaptic connectivity. Multiple parallel invaginated synapses are present in which the paired horizontal cell processes always lie lateral to one-to-several bipolar processes. Where the normal 18-30 nm spacing at chemical synapses is exceeded, as in the vertebrate receptor bipolar cell contact, it is possible that there is indirect transmission via another synaptic pathway (i.e., the horizontal cell processes). Divergent triad junctions in which a single terminal is presynaptic to three immediately adjacent postsynaptic processes are seen in both vertebrates amd arthropods. Reciprocal or feedback synapses are generally not well-characterized in vertebrate visual cells. Gap junctions have been reported between various vertebrate receptor cells using both freeze-fracture and thin sections for electron microscopy.

EVOLUTIONARY TRENDS

An ultrastructural comparison of invertebrate and vertebrate photo-receptor synapses suggests a tendency toward more complex synaptic structure in more highly evolved phyla. For example, synaptic vesicles become smaller and more numerous and more complex patterns of synaptic connectivity appear. There is even a morphological divergence of pre-synaptic components between arthropods and vertebrates (Fig. 8). Arthropods have developed a presynaptic bar that parallels the membrane whereas vertebrates have originated a presynaptic lamella that lies perpendicular to the membrane. Such elaborations of presynaptic components appear to be correlated with increased complexity in patterns of connectivity. Multiple contacts between similar and different second order neurons occur in a predominantly tiered pattern in arthropod retinula cells compared to horizontal arrangements of vertebrate rod and cone synaptic invaginations. Whether or not there are adaptive advantages to presynaptic bars and tiers of photoreceptor synapses in arthropods and presynaptic lamellae and horizontal rows of visual cell synapses in vertebrates remains to be evaluated. Bars or lamellae are also present in a few nonvisual cells such as in metathoracic ganglion neurons of the cockroach (31) and hair cells of the vertebrate acousticolateralis system (9). Their evolutionary significance needs to be established through further ultrastructural comparisons of presynaptic components in sensory receptors. We are left with the following questions: What evolutionary advantage does a membrane-separated presynaptic density offer two such divergent groups as arthropods and vertebrates? What interactions between variation of species and selection of pattern led to horizontal versus perpendicular orientation of presynaptic bars or lamellae? I can only point out that variation in synaptic structure appears where one presynaptic element serves more than one postsynaptic element.

Characteristics common to all groups are a close apposition of paired membranes with a uniform 18-30 nm space at chemical contacts and a 2-4 nm gap at electrical junctions. The uniformity of spacing at active sites is probably a structural limitation for chemical and electrical mechanisms of impulse transmission. That basic pattern preserves a recognizable thread of continuity throughout phyletic evolution even when accessory components are added. Because of the limited data available

I cannot begin to correlate what is going on at the synaptic end with the receptive end of visual cells. Whether or not synaptic patterns in visual cells support the mono-, di-, or polyphyletic lines of evolution discussed in the next three chapters remains to be seen. This survey of visual cell synapses, however, suggests an evolutionary divergence of photoreceptor presynaptic structure in those groups with the most sophisticated mechanisms of photoreception, namely, arthropods and vertebrates.

ACKNOWLEDGEMENTS

I thank the following for support: Robert D. Klemm, John C. Kinnamon, David E. Sims and grants NS 10264 and BNS 80-13061.

REFERENCES

1. Arnett-Kibel,C.,Meinertzhagen,I.A.,and Dowling,J.E.(1977): Proc. R. Soc. Lond. (Biol), 196:385-413.

2. Boschek,C.B.(1971): Z. Zellforsch. Mikrosk. Anat., 118:369-409.

3. Braitenberg,V.(1967): Exp. Brain Res., 3:271-298.

4. Burkhardt,W.,and Braitenberg,V.(1976): Cell Tissue Res., 173:287-308.

5. Cohen,A.I.(1973): J. Comp. Neurol., 147:379-398.

6. Custer,N.V.(1973): J. Comp. Neurol., 151:35-56.

7. Dowling,J.E.(1979): In: The Neurosciences, Fourth Study Program, edited by F.O. Schmitt and F. G. Worden, pp. 163-181. The MIT Press, Cambridge, Massachusetts, and London, England.

8. Dubin,M.W.(1974): In: The Eye, Comparative Physiology, Vol. 6, edited by H. Davson and L. T. Graham, Jr., pp. 227-256. Academic Press, New York and London.

9. Ginzberg,R.D.,and Gilula,N.B.(1980): J. Neurocytol., 9:405-424.

10. Gray,E.G.(1970): J. Cell Sci., 7:203-215.

11. Hafner,G.S.(1974): J. Neurocytol., 3:295-311.

12. Hámori,J.,and Horridge,G.A.(1966): J. Cell Sci., 1:257-270.

13. King,T.S.,and Dougherty,W.J.(1980): Am. J. Anat., 157:335-343.

14. Kolb,H.(1977): J. Neurocytol., 6:131-153.

15. Ladman,A.J.(1958): J. Biophys. Biochem. Cytol., 4:459-466.

16. Lasansky,A.(1967): J. Cell Biol., 33:365-383.

17. Melamed,J.,and Trujillo-Cenóz,O.(1966): Z. Zellforsch. Mikrosk. Anat., 74:12-31.

18.   Raviola,E.(1976): Invest. Ophthalmol., 15:881-895.

19.   Raviola,E.,and Gilula,N.B.(1973): Proc. Natl. Acad. Sci. USA,
      70:1677-1681.

20.   Ribi,W.A.(1978): Cell Tissue Res., 195:299-308.

21.   Shepherd,G.M.(1979): The Synaptic Organization of the Brain,
      2nd edition, pp. 192-195 Oxford University Press, New York, Oxford.

22.   Smelser,G.K.,Ozanics,V.,Rayborn,M.,and Sagun,D.(1974): Invest.
      Opthalmol., 13:340-361.

23.   Stell,W.K.(1972): In: Handbook of Sensory Physiology, Vol. VII/2,
      edited by M. G. F. Fuortes, pp. 111-213, Springer-Verlag, Berlin,
      Heidelberg, New York.

24.   Strausfeld,N.J.(1976): Atlas of an Insect Brain. Springer-Verlag,
      Berlin, Heidelberg, New York.

25.   Strausfeld,N.J., and Campos-Ortega,J.A.(1977): Science, 195:894-897.

26.   Theron,J.J.,Biagio,R.,Meyer,A.C.,and Boekkooi,S.(1979): Am. J. Anat.,
      154:151-162.

27.   Toh,Y.,Yoshida,M.,and Tateda,H.(1979): J. Ultrastruct. Res., 68:341-
      352.

28.   Trujillo-Cenóz,O.(1965): J. Ultrastruct. Res., 13:1-33.

29.   Trujillo-Cenóz,O.(1972): In: Handbook of Sensory Physiology, Vol.
      VII/2, edited by M. G. F. Fuortes, pp. 5-62. Springer-Verlag,Berlin,
      Heidelberg, New York.

30.   Vollrath,L.(1973): Z. Zellforsch. Mikrosk. Anat., 145:171-183.

31.   Wood,M.R.,Pfenninger,K.H.,and Cohen,M.J.(1977): Brain Res., 130:25-
      45.

32.   Yamasu,T.,and Yoshida,M.(1973): Publ. Seto Mar. Lab., 20:757-778.

33.   Yamasu,T.and Yoshida,M.(1976): Cell Tissue Res., 170:325-339.

Visual Cells in Evolution, edited by Jane A. Westfall, Raven Press, New York © 1982.

Continuity and Diversity in Photoreceptors

Richard M. Eakin

Department of Zoology, University of California, Berkeley, California 94720

Back when the world was young and life was new, Nature invented some molecules that permitted astounding evolutionary advances: chlorophyll that captured radiant energy for metabolism and photoreceptive proteins useful for phototaxis and later for vision. The new molecules, called photopigments, became embedded in membranes in a favorable orientation.

A conjectured first step in the evolution of photoreceptors might be an incorporation of a photopigment into the cell membrane (Fig. 1), suggested by a recent study of Melkonian and Robenek (28). They analyzed the eyespot membrane of an alga, <u>Chlamydomonas</u> <u>reinhardii</u>, using freeze-fracture technique. There is a unique cluster of particles on the protoplasmic face of the eyespot membrane in comparison with other regions.

FIG. 1. Unique particles in the area of the cell membrane (outlined in white) above the eyespot of an alga. From Melkonian and Robenek (28).

Particles measuring 3-12 nm are more numerous and particles 16-20 nm in
diameter less abundant in the plasmalemma over an eyespot than elsewhere
in the cell membrane. The smaller particles have approximately the same
size as the photoreceptive molecules in crayfish eyes (20) and in the eye
of a snail (1). Rhodopsin molecules, however, are smaller, being 4-5 nm
in diameter (35). An eyespot itself is a crescent of carotenoid droplets
that serves as a shading device, permitting the protist to determine the
direction of light and to respond by positive or negative phototaxis. In
another study (29), on released zoospores of the green alga Chlorosarcin-
opsis gelatinosa, Melkonian and Robenek found that the unique cluster of
particles on the protoplasmic face of the eyespot membrane disappears at
the time a zoospore settles and is no longer phototactic.

     The next step in the evolution of a photoreceptor might have been the
shift of light-sensitivity from the plasma membrane to the membrane of a
cilium or flagellum. At first, perhaps, only the base of a cilium possessed the photopigment. The modern protist Euglena spirogyra may serve
as a model for this evolutionary advance. A swelling at the base of one
of its flagella is the photoreceptor (21, 27). At my suggestion, Melkonian and Robenek analyzed the membrane over the flagellar swelling of
Euglena by freeze fracture. They found (personal communication) an aggregation of particles similar to that noted above in Chlamydomonas.

     Later the entire ciliary membrane presumably acquired a photopigment,
and the cilia thus endowed with light-sensitivity formed a group at some
strategic site on the surface of the organism. Then, to continue this
speculation, that unique spot invaginated to form a simple eye (Fig. 2).
The eyecups of some present day invertebrates may serve as models, such
as those of larval bryozoans studied by Woollacott and his associates

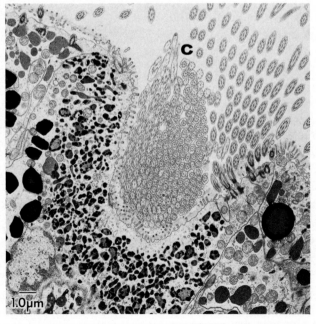

FIG. 2. Bryozoan ocellus with cilia (C), cut in cross section. From
Hughes and Woollacott (24).

(24, 39) or the ocellus of a larval endoproct (38). Hughes and Woollacott (25) recently concluded that only the horizontal cilia (from lateral cells) possess light-sensitivity. In Figure 2 they are seen in the
center of the ocellus. The vertical cilia (from posterior cells) are
thought to be nonphotoreceptive. In Figure 2 they arise from the margin
of the ocellus on the reader's right.

The light-sensitivity of a cilium could be increased by an expansion
in the area of the ciliary membrane, achieved by outfolding or infolding.
In the ocelli of certain hydromedusans (Phylum Cnidaria) there are cilia
with a large number of irregular microvilli that are outgrowths of the
ciliary membrane (13, 34, 41). Figure 3, of a remarkable ciliary photoreceptor from the marine hydromedusan Polyorchis penicillatus, illus-
trates this stage of evolution. Note in the inset that the axoneme is

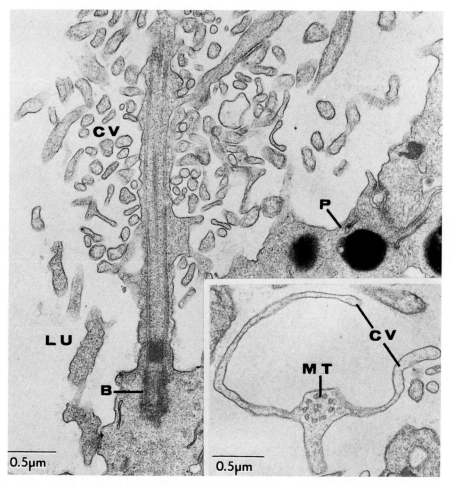

FIG. 3. Hydromedusan photoreceptor. Eakin and Brandenburger from Eakin
(6). B, basal body; CV, ciliary villi; LU, lumen of ocellus; P, pigmented cell. Inset: Cross section of ciliary photoreceptor. CV, ciliary
villi; MT, microtubules in an axoneme with a 9 x 2 + 2 pattern.

still a 9 x 2 + 2 pattern of microtubules. Later in the evolution of
ciliary photoreceptors the two central singlets are lost (9 x 2 + 0).

   Another way that the area of ciliary membrane was increased was by infoldings to form internal tubules that are exemplified by the ocelli of
arrowworms (2, 14). In the chaetognath _Sagitta_ _scrippsae_ the internal
tubules are very long (20 µm) but narrow (50 nm), lying in the distal
parts of the cilia (Fig. 4). Proximally the cilia contain many granules
whose function is unknown. We designated this part of a cilium a conical
body because of its shape (see lower inset of Fig. 4). The upper inset
shows a cross section of a cilium just above the basal body. Note that
the axoneme of the cilium has a 9 x 2 + 0 pattern.

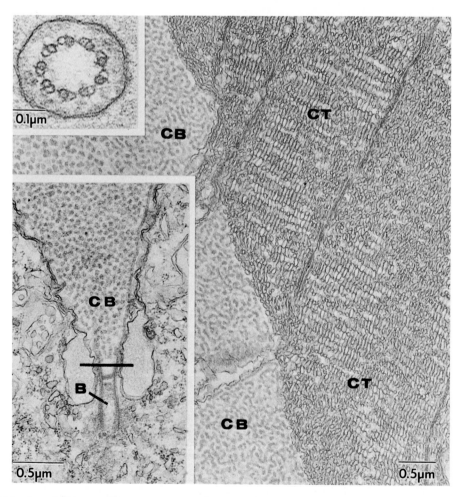

FIG. 4. Ciliary photoreceptors from a chaetognath. CB, conical bodies;
CT, ciliary tubules. Lower inset: Longitudinal section through base
of a cilium showing basal body (B) and conical body (CB). Upper inset:
Cross section of a photoreceptoral cilium taken at base of conical body
(see line in lower inset marking level). From Eakin and Westfall (14).

A third type of modification--vertically oriented lamellae formed by outfolding of the ciliary membrane--is exhibited by ocelli of ascidian larvae, such as that of Distaplia occidentalis (Fig. 5). The last instance of elaboration of a cilium as a photoreceptor to be cited here is the rod or cone of vertebrate lateral and pineal eyes, so well-known as to require no illustrations and references.

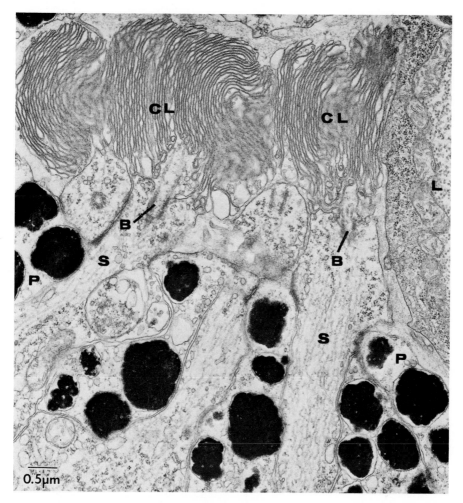

FIG. 5.   Photoreceptive cilia from eye of larval ascidian (Phylum Chordata).   From Eakin and Kuda (11).   B, basal bodies; CL, ciliary lamellae; L, lower border of lens; P, pigmented cell; S, sensory cells.

CILIARY LINE

I postulated (3, 4, 5) that from protist to vertebrate there is a ciliary line of evolution of photoreceptors (Fig. 6). The cilium, possessed of light-sensitivity by an incorporation of photopigment in its membrane, is the continuity. I doubt if many would advance a polyphyletic origin of cilia, considering the unique organization of their axonemes and basal

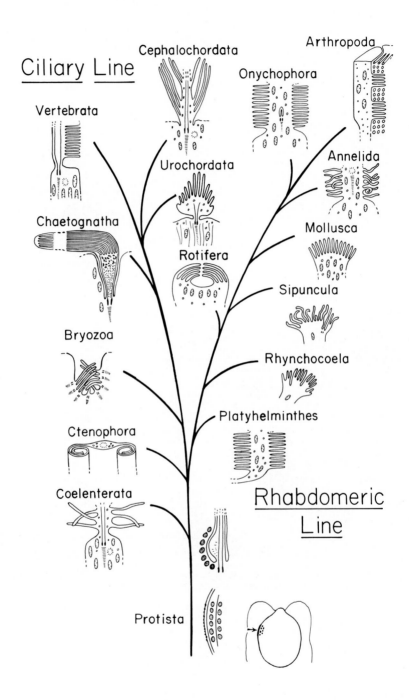

FIG. 6. Ciliary and rhabdomeric lines of evolution of photoreceptors.
Modified from Eakin (7) by deleting echinoderms and adding an alga.

bodies. <u>First</u>, it seems to me unlikely that such a complex organelle was invented separately and independently in the many animal phyla. A <u>second</u> premise in my theory follows: genetic coding for the synthesis and incorporation of a photopigment in ciliary membranes has also passed from ancestral group to ancestral group, the bases of the branches of the evolutionary tree. Point <u>three</u>: the various modifications of photosensitive cilia--the diversity--are due to mutations that occurred later, in the smaller branches or twiggery of the tree.

My colleague on this symposium program, Dr. L. v. Salvini-Plawen, and his associate, Professor Ernst Mayr, would probably accept my first premise but not the second one. They have interpreted Nature differently, postulating that "photoreceptors of various degrees of differentiation have been evolved independently in at least 40 if not 65 or even more separate phyletic lines" (33, p. 209), represented (p. 230) as a wheel, it seems to me, not as a tree. The hub is a generalized cell and the radiating spokes lead to the various types of photoreceptors, such as cilia with villi, lamellae, internal tubules, or fluted or whorled membranes. There are no <u>lines</u> of evolution in their theory.

It will be observed that I have revised Figure 6 (7) to delete the echinoderms from the ciliary line because recent studies (9, 40) reveal that photoreceptors in asteroids and in a holothurian are not ciliary. I believe, however, that the study of light-sensitive organelles in this phylum is incomplete. A ciliary photoreceptor may yet be found in some ancient group, such as the crinoids. Mrs. Brandenburger and I have made some unfruitful searches for evidence of a photoreceptor in two crinoids. A ciliary photoreceptor may also be discovered in some echinoderm larva. I believe that early developmental stages are more reliable for evolutionary speculation than adults. We have been unable to find a photoreceptor in a bipinnaria or echinopluteus larva. We hope to have an opportunity to study auricularia or doliolaria larvae. However, even if no example of a ciliary photoreceptor is found in a modern echinoderm, the ancestral stock of the phylum may have possessed this type of light sensor. According to Hyman (26, p. 696) "the original echinoderm must have been very different from those that we know today."

I do not propose, therefore, that the Echinodermata be transferred from the deuterostomes to the protostomes because villar photoreceptors occur on modern echinoderms. There are too many other features that the echinoderms share with most deuterostomes: enterocoelous coelom, radial cleavage, indeterminate and regulative development, stomodeal origin of the mouth, absence of chitin, creatine phosphate more abundant than arginine phosphate, <u>et</u> <u>cetera</u>.

RHABDOMERIC LINE

Further speculation: a second evolutionary line, rhabdomeric or microvillar, arose from the ciliary lineage by the formation of an array of villi or lamellae from the cell membrane. These elaborations of the plasmalemma contained opsins similar to those in the membranes of ciliary photoreceptors in having a chromatophore, retinaldehyde, and a protein carrier. I suggested that this innovation occurred below the origin of the flatworms (Phylum Platyhelminthes), as shown in Figure 6.

It is conceivable that at first an ocellus might have a mixture of ciliary and villar photoreceptors. Recently, Mrs. Brandenburger and I have found such an eye in the Müller's larva of a marine flatworm (10). We are indebted to Dr. Thurston Lacalli, University of Saskatchewan,

for directing us to this unique eye that was not completely analyzed by
Ruppert (32) in his study of metamorphosis of turbellarian larvae. The
left cerebral ocellus of the larva of <u>Pseudoceros canadensis</u> possesses
four sensory cells. Three of them bear microvilli and one a bundle of
about 50 cilia. The villi and cilia fill the cavity of the eyecup, composed of a single cell containing black (melanin ?) pigment granules. In
Figure 7 one sees these features and in addition the nucleus of the pigmented cell and the nuclei of the three villi-bearing cells. The nucleus
of the sensory cell with cilia is not shown. Both villi and cilia are
presumed to be light-sensitive because of their position and structural

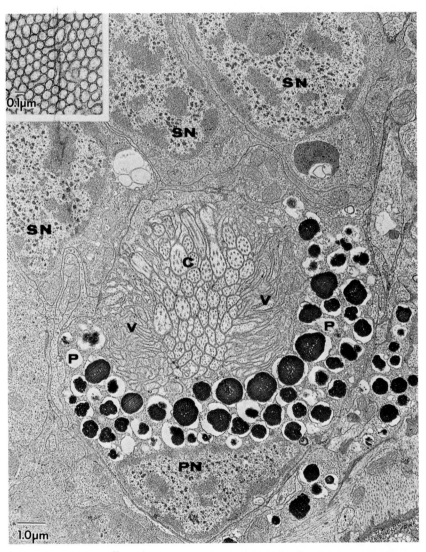

FIG. 7. Ocellus of Müller's larva of a marine flatworm. C, cilia; P,
pigmented cell; PN, pigmented cell nucleus; SN, sensory cell nuclei; V,
villi. Inset: Cross section of villi. Original, Eakin and Brandenburger.

similarity to those organelles in other eyes. The right ocellus has only three sensory cells, all of which possess villi. The significance of the asymmetry (we studied ten specimens) and the specific functions of rhabdomeres and cilia in the left eye are unknown. The cerebral ocelli of the adult, both left and right, are rhabdomeric.

   From the Platyhelminthes to the Arthropoda the rhabdomere is the dominant photoreceptoral organelle, although not without exceptions. In the majority of instances the rhabdomere consists of microvilli, but in some (e.g., the eye of a rotifer, Fig. 8) it is a stack of lamellae (15). The microvilli exhibit many different patterns. In some eyes they are irregular, as in the ocelli of many annelids; in other instances they are long and straight and orderly arranged, as in most arthropods, cephalo-

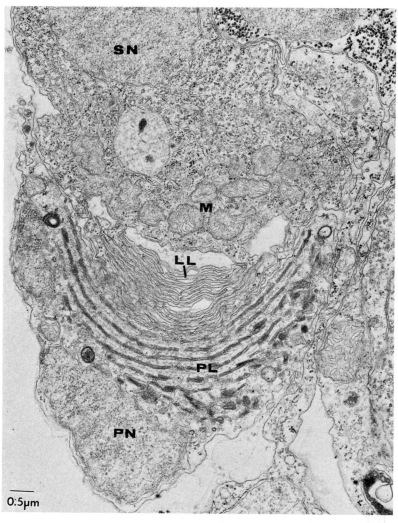

FIG. 8. Ocellus of a rotifer. From Eakin and Westfall (15). LL, stack of nonciliary lamellae; M, mitochondria; PL, pigmented plates; PN, pigmented cell nucleus; SN, sensory cell nucleus.

pods, onychophorans, and alciopid annelids (see 6 for references and
figures).

Among the annelids and mollusks there are many examples of ciliary
photoreceptors. In all instances, so far as presently known, they are
not in cerebral eyes. I have based my theory on brain eyes, believing
that they are more conservative and more reliable for phylogenetic speculation. Photoreception would have been their function throughout the
evolutionary histories of their possessors. Caudal, branchial, epidermal, tegmental or mantle eyes, being more peripheral and more subject to
environmental influences and selection, are more apt to exhibit cenogenetic changes. They evolved in tissues that were not originally photoreceptive. Either the coding for ciliary photoreceptors was retained in
the genome of the rhabdomeric line and given expression now and then in
non-cerebral organs or mutations established new genes for ciliary sensors. In either event, I hold that the rhabdomere is the principal, ongoing type of light receptor in the protostomes that culminated in the
Arthropoda in which ciliary photoreceptors are absent, so far as known.
I would apply the same argument to the situation in gastrotrichs, nema-
todes, nermertines, and flatworms where ciliary photoreceptors are epidermal or pericerebral. One example: the epidermal eye in the Müller's
larva, whose cerebral ocellus was described above, has unusual ciliary

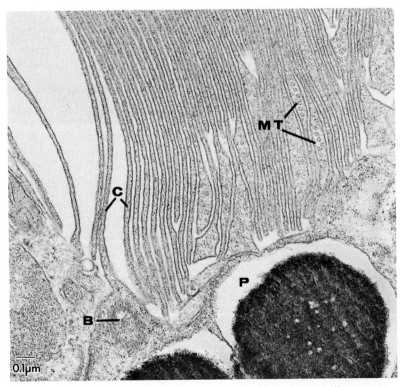

FIG. 9. Flattened cilia in the epidermal ocellus of a Müller's larva of
a flatworm <u>Pseudoceros</u> <u>canadensis</u>. Eakin and Brandenburger, original.
B, basal body; C, flattened cilia; MT, microtubules; P, pigmented cell.

receptors. They are lamellae (Fig. 9) arising from the pigmented cell that serves as a shading device and as the sensory cell. Our findings agree with those of Ruppert (32). There are other examples of ciliary photoreceptors in the Platyhelminthes (17, 18, 30).

Our study (12) of the ocelli of archiannelids was made in the hope that it might throw some light (figuratively) on the type of photorecep- tor in the ancestral annelids. I am well aware of the uncertainty that the archiannelids are primitive and that they may not be a monophyletic group (22, 31). At the time of our initial study we had representatives of four of the five families in the order. All had rhabdomeric eyes, and in only one (Nerilla antennata) were vestigial cilia found. We have now examined the ocellus of a member of the fifth family, the Polygordiidae. The photoreceptoral organelle is a rhabdomere, in agreement with that of the other archiannelids, thus strengthening the assumption that the an- cestral stock of the phylum Annelida possessed microvillar ocelli.

My colleague in this symposium, Dr. J. R. Vanfleteren, and his asso- ciate, A. Coomans, have advanced the hypothesis that all photoreceptors are basically ciliary (36). Dr. Westfall has called this proposition a monophyletic theory, in contrast to mine, a diphyletic one, and to that of Salvini-Plawen and Mayr (33), referred to above, a polyphyletic one. Vanfleteren and Coomans state in their paper (36, p. 165) that "the dif- ference between both receptor types [ciliary and rhabdomeric] is more of quantitative than of qualitative order. In both types the elaboration of the photoreceptoral organelle is induced by a ciliary formation that, after initiating membrane proliferation, may become more or less abortive (rhabdomeric type) or may develop further into a ciliary organelle (cil- iary type)."

The induction of microvilli by a cilium has not been established. I believe that because they are sometimes together in embryonic photosen- sory cells does not necessarily indicate a morphogenic relationship. One might ask: why are microvilli not induced in all instances of ciliated epithelium? Or, to turn the question around, why are cilia and basal bodies absent in the many instances of epithelial cells with microvillar (brush) borders? Eisen and Youssef (19) did not observe ciliary struc- tures in developing honeybee retinula cells. Home (23, for references) saw them in coccinellid and carabid beetles, but her observations do not convince me that a transitory ciliary bud, which may be noted sometimes when the microvilli first appear, induces a rhabdomere.

I can not rule out, of course, the possibility that some subtle induc- tive stimulus arises from a centriole and causes the formation of villi. It is easier for me to believe, however, that cilia in rhabdomeric eyes are adventitious. Ectoderm, from which cerebral eyes develop, is usually ciliated. Dr. Westfall, Mrs. Brandenburger and I have studied the devel- opment of two rhabdomeric eyes--in an onychophoran, Macroperipatus geayi (16), and in a snail, Helix aspersa (8). Basal bodies and ciliary buds were present at the apices of prospective sensory, pigmented supportive, and epidermal cells. The microvilli that formed at the distal surfaces of these cells did not appear to have an association with the ciliary structures. Typically there is a close relationship between an organizer and the induced structure. Wachmann and Hennig (37) made a study of the potential relationship between centrioles and the development of rhabdo- meres in a leaf-cutter bee. In the early pupal stage when a presumptive retinula cell is differentiating its rhabdomere, the centrioles lie in an end-to-end position below the nucleus. The authors concluded that "it was not possible to demonstrate any connexion between the centrioles and the developing microvilli of the rhabdomeres" (p. 337).

There are numerous examples of cilia formation from cytomembranes (not necessarily the plasmalemma) in cells not normally ciliated, as in those of certain glands: pituitary, adrenal, and pancreatic islets (of ecto-dermal origin) and in fibroblasts, chondrocytes, and osteocytes (of meso-dermal origin). For references see my earlier discussion (7).

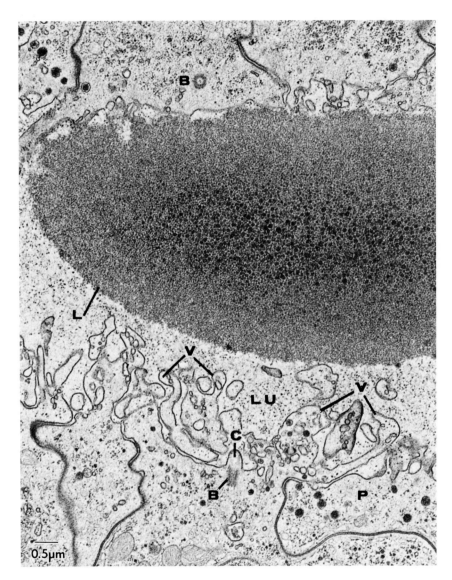

FIG. 10. Early stage in development of eye of the snail Helix aspersa. B, basal bodies (one above in prospective corneal cell, one below in a presumptive sensory cell); C, vestigial cilium in a developing sensory cell; L, newly formed lens; LU, lumen of ocellus; P, pigmented cell; V, young villi. From Eakin and Brandenburger (8).

CODA

The revelation by electron microscopists of the amazing variation in Nature at the ultrastructural level is like the finding of new species of animals and plants by naturalists. Both give the explorer the joy of discovery and the puzzle of the origins and evolutionary history of the diversities.

ACKNOWLEDGMENTS

The extent of my studies of photoreceptors, over twenty-five years, has been possible through the collaboration of many graduate students, research assistants and associates, and postdoctoral fellows. I name only two: Jane Westfall with whom I made early investigations and who has generously given time and effort as conceiver, organizer, editor and fund raiser of this symposium; and Jean Brandenburger, my research associate for sixteen years, who contributed to this paper by new electron-microscopy and preparation of the figures. I acknowledge also grants-in-aid (GM 10292, EY 02229, and GM 28778) from the National Institutes of Health and the personal encouragement of Dr. W. Sue Badman. A draft of the chapter has been read by Mrs. Brandenburger, Dr. Colin Hermans, Dr. Barbara Nichols, Dr. Ralph Smith, and Dr. Westfall. Finally, I am indebted to Dr. Margaret Franzen for the archiannelid specimens and to Dr. Robert L. Fernald and his associates for the Müller's larvae upon which new research was conducted for this paper.

REFERENCES

1. Brandenburger,J.L.,Eakin,R.M., and Reed,C.T.(1976): Vision Res., 16:1205-1210.

2. Ducret,F.(1978): Zoomorphologie, 91:201-215.

3. Eakin,R.M.(1963): In: General Physiology of Cell Specialization, edited by D. Mazia and A. Tyler, pp. 393-425. McGraw-Hill, New York.

4. Eakin,R.M.(1966): Cold Spring Harbor Symp. Quant. Biol., 30:363-370.

5. Eakin,R.M.(1968): In: Evolutionary Biology, Vol. 2, edited by T. Dobzhansky, M.K. Heckt, and W.C. Steere, pp. 194-242. Appleton-Century-Crofts, New York.

6. Eakin,R.M.(1972): In: Handbook of Sensory Physiology, Vol. VII/1, edited by H.J.A. Dartnall, pp. 625-684. Springer-Verlag, Berlin, Heidelberg, New York.

7. Eakin,R.M.(1979): Am. Zool., 19:647-653.

8. Eakin,R.M., and Brandenburger,J.L.(1967): J. Ultrastruct. Res., 18:391-421.

9. Eakin,R.M., and Brandenburger,J.L.(1979): Zoomorphologie, 92:191-200.

10. Eakin,R.M., and Brandenburger,J.L.(1981): Science, 211:1189-1190.

11. Eakin,R.M., and Kuda,A.,(1971): Z. Zellforsch. Mikrosk. Anat., 112:287-312.

12. Eakin,R.M.,Martin,G.G., and Reed,C.T.(1977): Zoomorphologie, 88:1-18.

13. Eakin,R.M., and Westfall,J.A.(1962): Proc. Natl. Acad. Sci. USA, 48:826-833.

14. Eakin,R.M., and Westfall,J.A.(1964): J. Cell Biol., 21:115-132.

15. Eakin,R.M., and Westfall,J.A.(1965): J. Ultrastruct. Res., 12:46-62.

16. Eakin,R.M., and Westfall,J.A.(1966): 6th Int. Congr. Electron Microsc. (Kyoto), 2:507-508.

17. Ehlers,B., and Ehlers,U.(1977): Zoomorphologie, 87:65-72.

18. Ehlers,B., and Ehlers,U.(1977): Zoomorphologie, 88:163-174.

19. Eisen,J.S., and Youssef,N.N.(1980): J. Ultrastruct. Res., 71:79-94.

20. Fernandez,H.R., and Nickel,E.E.(1976): J. Cell Biol., 69:721-732.

21. Grell,K.G.(1968): Protozoologie, 2nd ed. Springer,Berlin, Heidelberg, New York.

22. Hermans,C.O.(1969): Syst. Zool., 18:85-102.

23. Home,E.M.(1975): Tissue Cell, 7:703-722.

24. Hughes,R.L., and Woollacott,R.M.(1978): Zoomorphologie, 91:225-234.

25. Hughes,R.L., and Woollacott,R.M.(1980): Zool. Scripta, 9:129-138.

26. Hyman,L.H.(1955): The Invertebrates, Vol. 4, McGraw-Hill, New York, Toronto, London.

27. Leedale,G.G.,Meeuse,B.J.D., and Pringsheim,E.G.(1965): Arch. Mikrobiol., 50:68-102.

28. Melkonian,M., and Robenek,H.(1980): J. Ultrastruct. Res., 72:90-102.

29. Melkonian,M., and Robenek,H.(1980): Protoplasma, 104:129-140.

30. Palmberg,I.,Reuter,M., and Wikgren,M.(1980): Cell Tissue Res., 210:21-32.

31. Rieger,R.M.(1980): Zoomorphologie, 95:41-84.

32. Ruppert,E.E.(1978): In: Settlement and Metamorphosis of Marine Invertebrate Larvae, edited by F. Chia and M. E. Rice, pp. 65-81. Elsevier, New York.

33. Salvini-Plawen,L.v., and Mayr,E.(1977): In: Evolutionary Biology, Vol. 10, edited by M.K. Hecht, W.C. Steere, and B. Wallace, pp. 207-263. Plenum, New York.

34. Singla,C.L.(1974): Cell Tissue Res., 149:413-429.

35. Sjöstrand,F.S., and Kreman,M.(1978): J. Ultrastruct. Res., 65:195-226.

36. Vanfleteren,J.R., and Coomans,A.(1976): Z. Zool. Syst. Evolutions-forsch., 14:157-169.

37. Wachmann,E., and Hennig,A.(1974): Z. Morphol. Tiere, 77:337-344.

38. Woollacott,R.M., and Eakin,R.M.(1973): J. Ultrastruct. Res., 43:412-425.

39. Woollacott,R.M., and Zimmer,R.L.(1972): Z. Zellforsch. Mikrosk. Anat., 123:458-469.

40. Yamamoto,M., and Yoshida,M.(1978): Zoomorphologie, 90:1-17.

41. Yamasu,T., and Yoshida,M.(1976): Cell Tissue Res., 170:325-339.

Visual Cells in Evolution, edited by Jane A. Westfall,
Raven Press, New York © 1982.

A Monophyletic Line of Evolution? Ciliary Induced Photoreceptor Membranes

J. R. Vanfleteren

Instituut voor Dierkunde, Rijksuniversiteit Gent, Ledeganckstraat 35, B-9000 Gent, Belgium

Eakin (41,43,45,46,47) first focused our attention on the occurrence of two main types of photoreceptoral organelles: the ciliary type, derived from the ciliary membrane and the rhabdomeric type, which is derived from the distal cell membrane. In addition, there would be two main lines of photoreceptor evolution, a ciliary line in coelenterates and deuterostomes and a rhabdomeric line, considered as an offshoot of the former, typical for protostomes. Nobody, I think, shall underestimate the overwhelming impact of this provocative theory on our current knowledge of photoreceptor ultrastructure.

However, if we screen the various taxa we find that (1) some protostomous phyla possess ciliary photoreceptors (Gastrotricha, Kamptozoa, Bryozoa), whereas the deuterostomous Echinodermata and perhaps the Cephalochordata have only rhabdomeric ones; (2) that both types can be found in several phyla and often in the same animal; (3) that mixed receptors may occur; and (4) that many rhabdomeric receptors possess ciliary structures. We must conclude either that both ciliary and rhabdomeric photoreceptive organelles originated several times independently during metazoan evolution, at least as many times as there are unlinked exceptions to Eakin's theory, or that switching and mixing of types and the presence of characteristics of one type in the other, is indicative of a common ontogeny.

Thus Vanfleteren and Coomans (122) have suggested that the development of both ciliary and rhabdomeric photoreceptoral organelles are induced by a ciliary formation that may persist later (ciliary type) or may become abortive (rhabdomeric type). Consequently, structural differences between ciliary and rhabdomeric photoreceptor types would be more of quantitative than of qualitative order. Eakin (47), on the other hand, accepts the idea of independent evolution as the best explanation for the exceptions to his theory. Salvini-Plawen and Mayr (111) in going further along this thinking postulate that photoreceptors have originated independently in at least 40 if not 65 or more different phyletic lines and, consequently, reject Eakin's theory of diphyletic origin of photoreceptors.

Salvini-Plawen and Mayr (111) distinguish two more types of photoreceptoral organelles. The basic concept is that the organelles originating in (usually ciliated) epidermal cells are essentially different from those formed by already differentiated cells that have lost their potential of forming cilia, e.g., parenchymous and ganglionic cells. The

latter organelles are classified as a diverticular type, which is
phenotypically rhabdomeric, but of neuronal origin and essentially
devoid of ciliary structures. In addition to ciliary photoreceptive
organelles formed by infolding or outfolding of the ciliary membrane,
Salvini-Plawen and Mayr (111) distinguish those that achieve surface
enlargement by increasing the number of cilia and classify them as
unpleated ciliary type.

As does Eakin (47), I see no advantage to such distinctions but it
may be useful to refer to these types in an outline of photoreceptors
that occur throughout the animal kingdom.

PHOTORECEPTOR TYPES IN THE VARIOUS METAZOAN PHYLA

Cnidaria and Ctenophora

It is rather unusual for students of evolutionary pathways to share
common ideas about hypothetical ancestral forms. There seems to be
little doubt that the ancestral photoreceptive organelle evolved among
ciliated (flagellated) protists in close connection with a cilium
(flagellum; 69,133). Most probably shading pigment was incorporated
also at an early stage of evolution.

The photoreceptor cells of the scyphomedusan Cassiopeia (12) may
exhibit such a primitive condition in that they also contain pigment,
but in the other cnidarians studied there are separate photoreceptor and
pigment cells. In Leuckartiara (116) the receptive cilia do not possess
membrane evaginations, whereas in other medusae, such as Polyorchis (54),
Bougainvillea (116) and Cladonema (12), the cilia form several micro-
villous processes (Fig. 1A-D). Lamellar extensions of the ciliary membrane have been reported for the hydromedusa Phialidium (12) and the
ctenophore Pleurobrachia (79,91). However, in Phialidium the cilia
retain the characteristic configuration of motile cilia (9+2 axonemes),
whereas in Pleurobrachia, the axoneme structure is 9+0.

Platyhelminthes

In adult Turbellaria the ocelli are usually situated in the parenchym
of the head region and are of the pigment cup type (Fig. 1E). The
predominant receptor type is the diverticular type of Salvini-Plawen
(6,7,19,20,39,60,61,90,97,98,105,118).

Ciliary receptors with supposed photoreceptive function have recently
been described in the head region of Proseriata (Fig. 1F). In
Parotoplanina (60) the putative photoreceptive organelle is a lamellate
body, derived from about 10 rootless cilia with 9+0 microtubules. In
Dicoelandropora and Notocaryoplanella (61) there are ciliary aggregations of numerous rootless cilia (with 9+0 axonemes at their bases);
both are good examples of the unpleated type of Salvini-Plawen and
Mayr (111).

Ruppert (110) studied the eyes in young Müller's and Götte's larvae
of Turbellaria. Both had paired cerebral eyes and a single epidermal
eye, the fine structure of the latter being identical in both types of
larvae. The sensory cell contains pigment as well as several cilia
modified to form flattened membranes oriented perpendicularly to the
direction of incident light. This cell is covered by a second cell,
whose nucleus supposedly functions as a lens (Fig. 2C). According to
Ruppert (110), the cerebral eyes consist of a pigment cell, ciliated

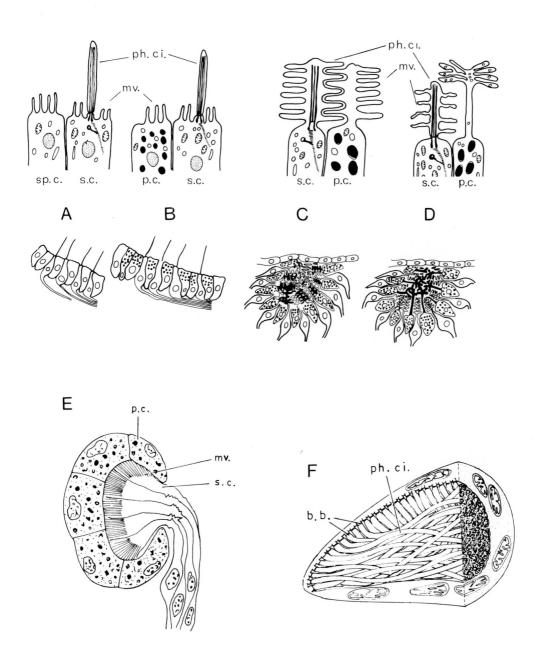

FIG. 1. A–D:Morphological sequence of photoreceptor differentiation in
Cnidaria. A:Hypothetical indistinct ocellus. B:Ocellus of Leuckartiara.
C:Ocellus of Polyorchis. D:Ocellus of Bougainvillea. After Singla
(116). E:Inverted pigment cup ocellus of a planarian. After Carpenter
et al. (20). F:Ciliary photoreceptor of Proseriata. After Ehlers and
Ehlers (61).

cells, and cells producing a rhabdomeric layer of microvilli. In the cerebral eye of Müller's larva, the rhabdomere is well developed with few cilia present (Fig. 2A), whereas in Götte's larva abundant cilia form a bulb adjacent to the smaller rhabdomeric layer (Fig. 2B). In a recent study, however, Lanfranchi et al. (94) reported no difference in the cerebral eyes of Müller's and Götte's larva of polyclad Turbellaria. According to these authors, the cerebral eyes are made up of three cells only: one pigmented cell, which gives rise to the bulb of cilia, and two rhabdomeric sensory cells of ganglionic origin. Ruppert (110) assumed the bundle of cilia to function as a lens, whereas Lanfranchi et al. (94) ascribed a photoreceptive function to the cilia.

The ocelli of larval Trematoda are predominantly rhabdomeric (16,66, 67,82,84,96,104) but ciliary receptors with presumed photoreceptive function have been found in the head of the oncomiracidium of Entobdella (96) and in the miracidia of Diplostomum and Schistosoma (113). They resemble the structure described above for Parotoplanina, except for different axoneme arrangements:9+2 in Entobdella (Fig. 2D), 9+0 in miracidia of Diplostomum and Schistosoma and 9+0 (rarely 8+0 or 8+2) in cercariae of Schistosoma.

Nemertini

Nemerteans may possess ocelli as well as cerebral photoreceptive organs. The ocellar photoreceptors are rhabdomeric, but ciliary formations, e.g., a well developed rootlet may be present (119,125). The cerebral photoreceptors, which one might expect to belong to Salvini-Plawen and Mayr's (111) diverticular type, are clearly ciliary formations. They exhibit basal bodies and 9+0 arrangements of microtubules (126; Fig. 3A).

Gastrotricha

Gastrotrichs may possess photoreceptor cells of presumed ganglionic origin (111), the fine structure of which to date has only been studied in Turbanella (121). Here, the presumed photoreceptive organelles are composed of numerous microvilli, which originate from a modified cilium with basal body and rootlet; consequently they belong to the ciliary type (Fig. 3B).

Nematoda

The nematode ocelli studied are predominantly rhabdomeric and entirely devoid of ciliary structures (27,28,114,115). In Oncholaimus, however, one of the amphidial dendrites forms a photoreceptive organelle of the unpleated ciliary type. Basal bodies, rootlets and arms on the a-tubules are lacking and the axoneme arrangement is variable:7-9 doublets and 0-3 central singlets (17,18).

Rotatoria

Rotifers possess anterior ocelli in apical or lateral position and cerebral photoreceptors embedded within the brain. Salvini-Plawen and Mayr (111) regard the former as epidermal receptors, the latter as derived from ganglionic cells.

In the cerebral eyes of Asplanchna (57), Trichocera, Brachionus, and

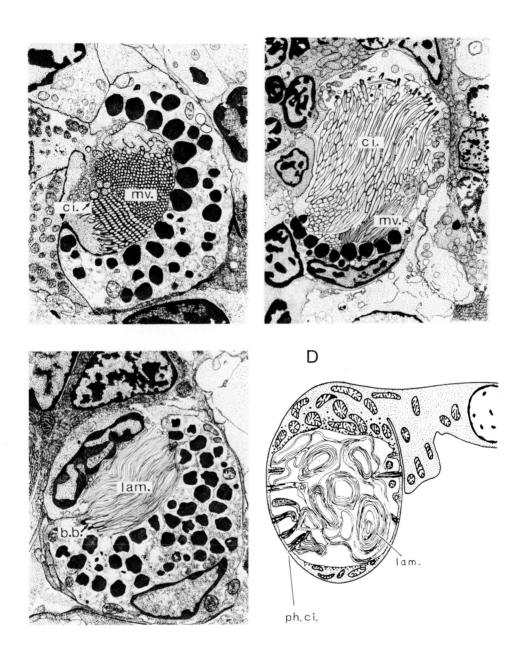

FIG. 2. A–C:Photoreceptors of turbellarian (Polycladida) larvae. After
Ruppert (110). A:Cerebral eye of Müller's larva. B:Cerebral eye of
Götte's larva. C:Epidermal eye of Müller's larva. D:Ciliary receptor
of <u>Entobdella</u>. After Lyons (96).

Rhinoglena (23-25) the receptive organelles represent the ganglionic diverticular type of Salvini-Plawen and Mayr (111), but in the cerebral eyes of Philodina (25) the putative photoreceptive organelles are ampullae-shaped cilia that contain electron dense material (Fig. 3C-E). Whereas the shape of these cilia is atypical for the photoreceptive function, the presence of pigment in surrounding epithelial cells is rather indicative of such a function. Moreover, the anterior ocelli of Trichocerca similarly contain ampullae-shaped cilia, whose contents are dense (Fig. 4A). The anterior receptor of Philodina is also of the ciliary type, but here the cilia bear lateral lamellar expansions that are piled up (Fig. 4B). Sensory organelles of the anterior ocelli of Rhinoglena are piled dendritic lamellae devoid of ciliary structures (25).

Mollusca

Placophora

Chitons possess so-called shell eyes that have differentiated from aesthetes (dorsal nerve strands penetrating the shell plates). In the intrapigmental shell eyes of Chiton, distal parts of the sensory cells penetrate the pigment cup and form numerous long villi and a few cilia in it (70). The extrapigmental ocelli of Onitochiton possess both rhabdomeric and ciliary sense cells. The rhabdome in the center of the eye is formed by surrounding cells, which also bear a few short cilia with 9+2 axonemes and short rootlets. Lamellate bodies derived from ciliary (9+2 axonemes) membranes are formed by cells at the periphery of the eye cup (13). Both types of photoreceptive cells are associated with different aesthetes in Acantochiton. Here rhabdomeric visual cells are situated in the median aesthetes; ciliary lamellate cells with supposed photoreceptive function are linked to the much smaller aesthetes in the lateral fields (65).

The trochophora larva of the black chiton Katharina tunicata bears a pair of ocelli consisting of several pigmented cells and one sensory cell, the distal end of which gives rise to a disorderly array of microvilli and a cilium with 9+2 arrangement of microtubules (109). The whole structure is very similar to that described for the sensory cells of the sipunculid Phascolosoma.

Gastropoda

The photoreceptive organs of Gastropoda are predominantly situated on the head. The sensory cells are of the rhabdomeric type and possess well developed, reduced, or no cilia (14,21,41,42,48,51,59,80,107,112). The eyes of Aplysia (80) are peculiar. Here, sensory cells with a few cilia and numerous microvilli alternate with cells carrying about equal numbers of cilia and microvilli and that conceivably are also photoreceptive. All cilia have 9+2 axonemes and basal bodies, but lack rootlets (Fig. 4C). Onchidium possesses dorsal eyes with ciliary (9+0 axoneme arrangement) lamellate sensory cells in addition to the rhabdomeric tentacular eyes (139). Finally, Dilly (33) reported on disks of ciliary origin in the sensory cells of the telescopic eyes of Pterotrochea, but his interpretation has been questioned (46).

Bivalvia

In Bivalvia there are cerebral eyes and/or pallial eyes. The cerebral eyes of Mytilus are of the rhabdomeric type. Each sensory cell carries a single cilium with 9+2 axoneme in addition to numerous microvilli (108). Pallial eyes have been studied mainly in Arcaceae,

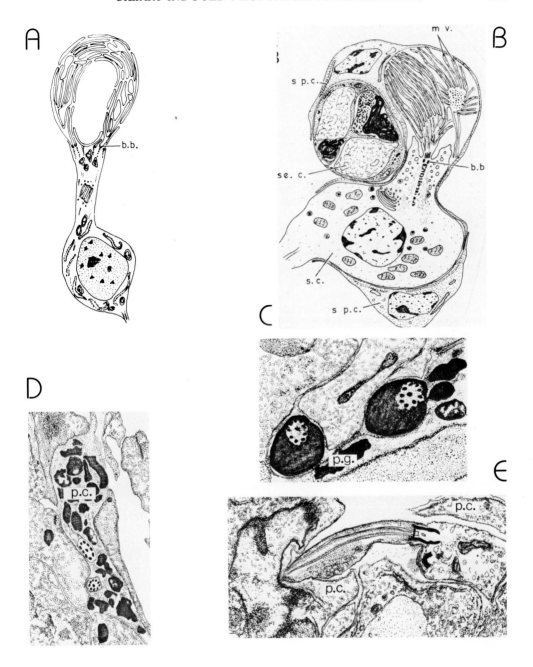

FIG. 3. A:Cerebral photoreceptive cell of Lineus. After Vernet (126).
B:Anterior cephalic sense organ of Turbanella. After Teuchert (121).
C-E:Cerebral eye of Philodina. C:Transverse section of both photoreceptive cilia. D:Transverse section of their bases. E:Axial section
of one of these cilia. After Clément (25).

Pectinaceae and Cardiaceae. In Arca and Glycymeris the light sensitive organelles consist of closely packed membranous disks that originate from cilia with 9+2 axonemes at their bases but no rootlets (95). Visual cells, situated at the tips of eye-bearing tentacles of Cardium carry numerous cilia that may become flattened and form lamellae. These modified cilia have basal 9+0 axonemes and basal bodies but no rootlets (3). The pallial eyes of Pecten are peculiar in having a double retina. The sensory cells of the distal retina (facing the lens) bear lamellate organelles at their distal surfaces, derived from modified cilia with 9+0 azonemes at their bases, but no rootlets. The receptor surfaces of the proximal sensory cells are directed opposite to those of the distal cells. They are composed of numerous microvilli and one or two cilia with 9+0 axonemes and conceivably belong to the rhabdomeric type (2).

Salvini-Plawen and Mayr (111) propose a morphological sequence of differentiation in the mantle eyes of Bivalvia from the single everse pit of Arca toward the closed lens eye with inverted retinal cells and a ciliary sense organ in Cardium and the closed lens eye of Pecten with proximal retina as in Cardium and ciliary sense organ representing the distal retina. Though supported by anatomical evidence, this elegant hypothesis not only involves replacement of one sensory cell type by another, but also substitution of one sensory function by another, because the sense organ of Cardium presumably has an olfactory function (3).

Cephalopoda

Finally, the very complex eyes of the Cephalopoda are of the rhabdomeric type and completely lack ciliary structures (Fig. 4D). No information is available about their early developmental stages, however (129,137,140,141).

Sipunculida

Akesson (1) distinguishes three different types of ocelli within sipunculids: larval, cerebral and tentacular. Of these, only the cerebral pigment spot ocelli of Phascolosoma have been investigated ultrastructurally. The visual cells bear numerous microvilli at the expanded apex of each cell and at least one cilium (usually 9+2 or rarely 8+2 axonemes without rootlets) with a shaft invaginated into the cell body. The microvilli appear to arise from the apical cell membrane without connection to the ciliary membrane; hence, this receptor is considered to be of the rhabdomeric type (73,74).

Annelida

Polychaeta

These worms possess cephalic photoreceptors that may be either inverse pigment cup ocelli or everse epidermal eyes. The pigment cup ocelli are of the diverticular type of Salvini-Plawen and Mayr (111): they are devoid of ciliary structures. The larval ocelli reportedly also belong to this type. They may persist in the adult or be replaced later by the everse epidermal eyes (111). The latter may contain ciliary rudiments. In addition to the cephalic eyes, special photoreceptive organs may differentiate in other body regions, e.g., metameric, pygidial and tentacular eyes.

The Archiannelida possess cephalic ocelli. In Protodrilus and

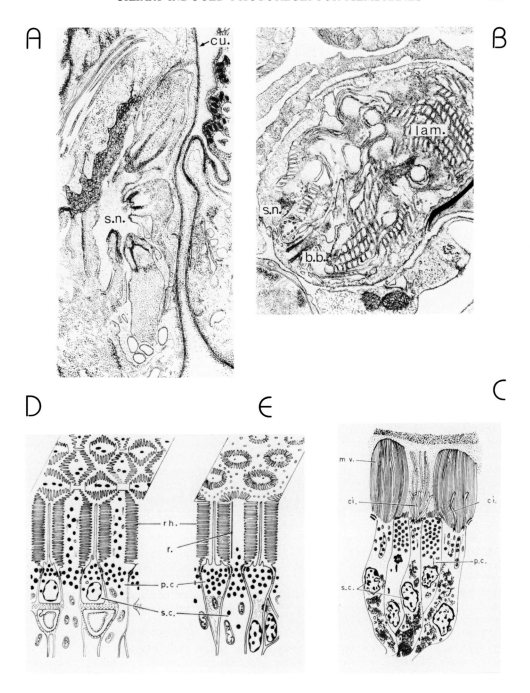

FIG. 4. A:Anterior sense organ of <u>Trichocerca</u>. B:Anterior receptor of
<u>Philodina</u>. A and B after Clément (25). C:Retinal cells of <u>Aplysia</u>.
After Hughes (80). D:Detail of the retina of <u>Octopus</u>. After Yamamoto
et al. (137) and Hermans and Eakin (75). E:Primary retina of <u>Vanadis</u>.
After Hermans and Eakin (75).

Saccocirrus these are inverse, rhabdomeric and contain no ciliary structures. The ocelli of Dinophilus are everse, but still rhabdomeric and devoid of cilia. The latter are reported from Nerilla only, but are apparently not connected with the rhabdomeres (52,99).

The prostomial (=cephalic) eyes of Phyllodocidae have been studied in Eulalia by Whittle and Golding (132). A single pair of compound everse eyes is situated adjacent to the anterio-lateral corners of the brain, immediately beneath the epidermis. The fine structure of the sensory cell is interesting in that there is a distal segment with an array of lateral microvilli and occasionally whorled membranes and a core with a prominent striated root arising from an apical centriole. The plasma membrane above the centriole often forms a bulbous protrusion, but a ciliary axoneme has not been found (Fig. 8B). Sensory cells, similar in cytology, but lacking a centriole, occur in the back of the brain. They are associated with few pigment cells and have been called posterior photoreceptors by Whittle and Golding (132). Apart from these rhabdomeric receptors, the same authors (132) describe about ten ciliary receptors with presumed photoreceptive function, situated immediately behind the complex eyes. The terminal parts of the sensory cells consist of several modified cilia and a number of microvilli. The cilia (9+0 axonemes with basal bodies but no rootlets) give rise to several branches, each usually with a single doublet.

The visual cells in the prostomial eyes of Arctonoë (Polynoidae) are very similar to their relatives in Eulalia, except that the apical centriole may be missing (117).

The ultrastructure of the elaborate prostomial eyes of the Alciopidae has been studied by Hermans and Eakin (75). These eyes are equipped with one main and one or more accessory retinas, whose sensory cells are essentially similar. They are of the rhabdomeric type with densely packed microvilli on all sides of a distal process. Cilia are absent, but a prominent root runs along the process and is connected to a distally situated basal body (Fig. 4E). Not only the visual cells, but the whole complex eyes of alciopids and cephalopods are markedly similar, an excellent example of parallel evolution.

Syllidae may reproduce asexually by budding. The posterior individual (bud) then generates a head before or after it is released from the parent organism. The fine structure of the photoreceptive cells of the parent organism, which developed from a larva, and of the budded organism is very similar in Syllis. Each sensory cell possesses a distal process bearing a rhabdomere of densely packed microvilli and a rudimentary cilium with 9+0 axoneme, basal body and prominent root running along the process. This situation is very similar to that found in Alciopidae and Phyllodocidae, except for the presence of the rudimentary cilium itself (10,11).

The ontogeny of these eyes has been studied in Syllis (11) and in Autolytus (124). From those results it appears that the modified cilium is present in the very early stages of differentiation. Microvilli are believed to develop independently at the base of the cilium, which is lifted up by simultaneous differentiation of the apical process, but retains its connection to the neck of the cell through its root. The sensory cells of Odontosyllis (9) and Autolytus (8) are very similar to those of Syllis.

In Nereidae, both larval and prostomial ocelli have been studied. The inverse ocelli of the trochophora larva of Neanthes consist of one pigment cell and one sensory cell of the rhabdomeric type, without a cilium, but with a centriole and a striated rootlet (55). The sensory cells of juvenile or developing everse prostomial eyes are likewise rhabdomeric and may possess ciliary structures, e.g., a cilium and a basal body in <u>Neanthes</u> (55) and a basal body in <u>Platynereis</u> (64). Striated rootlet, basal body and perhaps a very rudimentary cilium are retained in the adult sensory cells of <u>Platynereis</u> (64), but a prominent root only has been found in those of <u>Nereis</u> (35; Fig. 8E).

In addition to the prostomial eyes, ciliary sense cells with presumed photoreceptive function are embedded in the brain of <u>Nereis</u> (29). They bear modified cilia (9+0 axonemes) that give rise to several branches, thus strongly resembling the ciliary photoreceptors of <u>Eulalia</u> described above.

The Nephtyidae possess paired pigmented ocelli embedded in the brain and a pair of accessory photoreceptive organs on each side of the prostomium with identical sensory cells. These are of the rhabdomeric type and devoid of ciliary structures (142). In addition to the former photoreceptive organs, Zahid and Golding (142) described a pair of "sensory sacs". These are also embedded in the brain and consist of two cells enclosing a vacuole (phaosome) with microvilli and, in one case, a cilium projecting from the limiting membrane. These cells are supposed to be sensitive to light because of their similarity with phaosome photoreceptors of Hirudinea and Oligochaeta (see below; Fig. 5B).

<u>Armandia</u> (Opheliidae) possesses three prostomial ocelli and two groups of photoreceptor-like cells that are all embedded within the brain. No ciliary structures have been found in the sensory cells (71,72). Likewise rhabdomeric are the segmental ocelli, which are situated in front of the parapodia of the 7th to the 17th segments. The lateral ocelli are generally considered to be secondary formations because they develop as strictly adult specializations after settling and metamorphosis of the larva (71).

A quite different type of newly formed ocelli is found in the Sabellidae on the branchial tentacles. It has been studied in <u>Branchiomma</u> (92), <u>Dasychone</u> (85,86), <u>Potamilla</u> (87), <u>Sabella</u>, <u>Eudistylia</u>, <u>Pseudopotamilla</u> and <u>Schizobranchia</u> (88). The sensory organelles are invariably lamellar sacs that are extensions of the membranes of modified cilia (9+0 axonemes) without rootlets. There is, however, more diversity in the arrangement of the sacs within the sense cells (Fig. 5A).

Clitellata

These possess photoreceptor cells of the (rhabdomeric) phaosome type (Fig. 5B), which may occur as single cells or may be aggregated into ocelli. Ciliary structures may be present in the phaosome, e.g., some cilia with 9+0 axonemes in <u>Lumbricus</u> (106) and centrioles in <u>Helobdella</u> (22). White and Walther (131) have demonstrated that the phaosome membrane is continuous with the plasma membrane in <u>Hirudo</u>.

Pogonophora

Phaosomes have also been found in two clusters of presumed photoreceptor cells in the front end of <u>Siboglinum</u>. They are similar to those found in hirudineans, but ciliary structures are lacking (102).

Onychophora

<u>Peripatus</u> possesses two well developed eyes at the base of the antennae. The photoreceptive cells are of the rhabdomeric type with

numerous microvilli and a rudimentary cilium (9+0 axoneme, basal body
and a poorly differentiated rootlet) enclosed in an extracellular space
at the base of the sensory process (58). An ontogenetic study of the
eye of Macroperipatus showed that the photoreceptive cells develop from
ciliated ectodermal cells, the distal parts of which differentiate into
a microvillar process that overgrows the modified cilium (43).

Arthropoda

The photoreceptive cells in Crustacea and Insecta are exclusively
rhabdomeric whether they occur in simple ocelli, stemmata or in the very
elaborate compound eyes. The characteristic sensory apparatus of the
latter is the rhabdome that is formed by the rhabdomeres (consisting
of straight microvilli) projecting from the sides of several retinula
cells.

Ciliary structures have long been unknown in the receptor cells of
arthropods, but in 1972 Home (76) first reported the presence of
centrioles and rootlets in the retinula cells of Ladybird beetles
(Coccinellidae). Each retinula cell contains, at its distal end, a pair
of centrioles aligned in tandem, from which a ciliary root extends
proximally and runs alongside the rhabdomere (Figs. 5C and 8D). Thus
as in many annelid eyes it appears to be the ciliary root that is
associated with the rhabdomeric microvilli. An ontogenetic study of the
coccinellid photoreceptor revealed that tandem centrioles with rootlets
are already present at the time of pupation. The distal centriole
subsequently becomes the basal body of a transitory ciliary bud, whose
formation precedes the onset of rhabdomere formation (77). Paired
centrioles and associated ciliary roots also occur in the adult retinula
cells of carabid and cincindelid beetle eyes (78,93).

Centrioles have been found in juveniles of Megachile (Hymenoptera),
Carausius and Locusta (Orthoptera). In Carausius and Locusta they lie
close to the developing rhabdomere (120,128). Wachmann and Hennig (127)
examined one stage of pupal eye development in Megachile, a stage at
which rhabdomere formation had already begun. Here the centrioles were
situated at a considerable distance from the rhabdomere. On the other
hand, Juberthie and Munoz-Cuevas (83) observed centrioles and ciliary
microtubules in close contact with the growing microvilli of the visual
cells in the embryo of Ischryopsalis (Opiliones), and Munoz-Cuevas (101)
proposed a ciliary model for rhabdomere development in this arachnid.
Finally, Eisen and Youssef (62) found parallel paired centrioles close
to the developing rhabdomere in an early stage (fifth instar larva) of
eye development in the worker honey bee. It is not known whether they
were also present in preceding stages or occurred de novo in the fifth
instar. They may disintegrate later one, since they have not been
observed in subsequent stages nor reported in the eyes of adult workers
on drones.

Lophophorata

The larval photoreceptors of both Kamptozoa and Bryozoa belong to the
unpleated type of Salvini-Plawen and Mayr (111), each sensory cell
bearing a tuft of tightly packed cilia. These appear to be unmodified
kinocilia (9+2 axonemes and ATPase arms on the a-tubules) in the bryozoans Bugula (136) and Scrupocellaria (81; Fig. 5D) and only lack
ATPase arms in kamptozoan larvae (135).

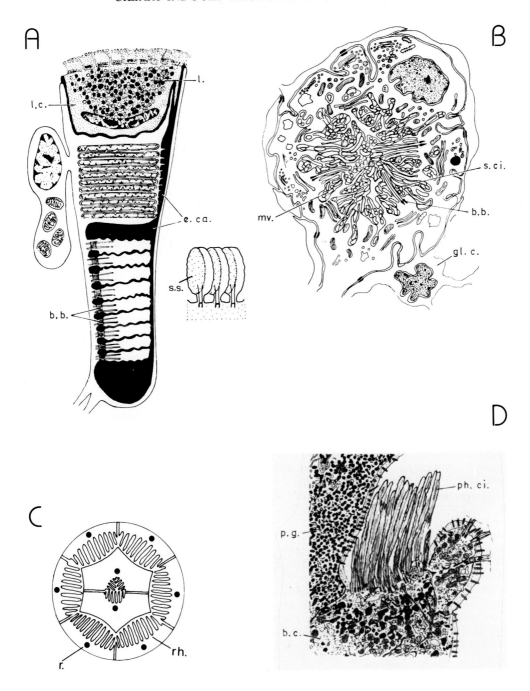

FIG. 5. A:Receptor and lens cells in the eyespot of Branchiomma and detail of sensory sacs. After Krasne and Lawrence (92) and Salvini-Plawen and Mayr (111). B:Phaosome of Lumbricus. After Rohrlich et al. (106). C:Retinula of Adalia. After Home (76). D:Ocellus of Scrupocellaria. After Hughes and Woollacott (81).

Echinodermata

Many Asteroidea possess everse ocelli in their arm tips. Earlier ultrastructural studies performed on the sea stars Henricia, Leptasterias, Patiria (41,45) and Asterias (123) showed the ocellar cavities to be filled with sensory cell processes, microvilli and cilia with 9+2 or 9+0 axonemes, but it remained unclear whether the microvilli emerged from the bases of the cilia (41,45) or arose independently from the apical cell surface (123). The latter arrangement certainly occurs in the ocelli of Opheodesoma (Holothuria). Here, the microvilli are not formed by modification of the ciliary sheath, but arise as outgrowths of the plasma membrane. Moreover, the effects of light and darkness on the holothurian ocellus unmistakebly proved that the microvilli are the light sensitive organelles (138). A restudy of the asteroid ocellus, undertaken by Eakin and Brandenburger (49) in Leptasterias, Henricia and Patiria then led to the final conclusion that here also microvilli are the photoreceptive organelles; they arise from the sensory processes without any clear morphological connection to the cilia. This interpretation is also compatible with a recent report on the fine structure of the optic cushion in the asteroid Nepanthia (103). Thus the echinoderm photoreceptor must be assigned to the rhabdomeric category.

Chaetognatha

Arrowworms possess a pair of inverse (Sagitta, Serratosagitta) or everse (Eukrohnia) eyes, situated posteriorly on the dorsal surface of the head. The photoreceptor cells bear highly aberrant cilia with a proximal 9+0 arrangement of microtubules connected to a basal body and a short cilium or long rootlet (36,37,56). In Sagitta (Fig. 6A), the sensory cilium is partly invaginated into the cell body. It readily expands above its base, forming a conical body surmounted by a distal tubular segment, which is presumed to be formed by infolding of the ciliary membrane. The proximal and distal parts of the conical body contain granules and short irregular cords respectively. A crown of microvilli arising from the distal part of the sensory cell encircles the upper part of the conical body (56). In Serratosagitta (Fig. 6B) the conical body has a petaloid structure with nine "leaves". The apical microvilli of the sensory cell are much longer than those in Sagitta. They surround the tubular segment and invaginate into its membrane (37). In visual cells of Eukrohnia (Fig. 6C) the conical body differentiates three zones of which the basal one is granular, the middle area is filled with short, then longer cords and the distal zone is opaque and may possibly function as a lens. Densely packed microvilli, which presumably arise from the ciliary membrane, surround the middle part of the conical body and are lodged in excavations of the former (37).

Hemichordata

Tornaria larvae of the enteropneust Ptychodera have everse ocelli, each sensory cell of which has a bulbous cilium (9+2 axoneme, basal body and striated rootlet) at its distal end and one, sometimes two arrays of microvilli from its sides below the cilium (Fig. 6D). Unfortunately, it is as yet unknown whether cilia or microvilli or both are the light-sensitive organelles. It is possible that tornarian photoreceptors may be of a mixed type (15).

Tunicata

Tadpole larvae of Ascidiaceae typically possess in their cerebral vesicle one ocellus, the fine structure of which has been studied in Ciona, Distalpia and Amaroucium (4,5,30,31,50). Each sense cell bears a distal process at the tip of which a cilium (9+0 axoneme, basal body and ciliary rootlet) gives rise to numerous membranous lamellae (Fig. 6E). The ontogeny of this receptor has been studied in Amaroucium (5). First a balloon-shaped evagination of the cell membrane is formed above the ciliary basal body, then the shaft of the cilium develops and microvilli arise from the cell membrane adjacent to the cilium. The latter increases greatly in diameter and its basal membrane envelops the surrounding microvilli. Next, the ciliary membrane infolds to form many finger-like projections closely intertwined with the microvilli. The projections soon assume a lamellar nature and extend beyond the microvilli both distally and laterally (Fig. 9B). This receptor is generally accepted as belonging to the ciliary type, although it remains to be proven that light sensitive molecules are associated exclusively with the ciliary membranes. A second type of photoreceptor cell, described by Dilly (32), is more likely to be a pressure receptor (50). Still other sensory cells are situated in pigment spots around the siphons of adult Ciona. Each bears a cilium and several microvilli that are apparently not associated with the cilium (34). The photoreceptive function of these cells has been questioned (134); if they are sensitive to light, they are to be classified with the rhabdomeric type.

The sense cells in the inverse ocelli of Salpa (Thaliaceae) belong to the ganglionic type of Salvini-Plawen and Mayr (111). The light sensitive organelles consist of an irregular array of microvilli with no trace of ciliary formations. Thaliacean photoreceptors are secondarily differentiated parts of the nervous system (68).

The photoreceptor of Salpa is exceptional in generating a hyperpolarizing response to light. All rhabdomeric type receptors recorded so far produce depolarizing receptor potentials as a response to illumination, whereas the ciliary ones are hyperpolarizing (68).

Cephalochordata

Branchiostoma (Amphioxus) possesses inverted pigment cup ocelli (Hesse cells) that have a ganglionic origin and are regarded as secondarily acquired differentiations. Thus by definition the receptor cells belong to the ganglionic-diverticular type of Salvini-Plawen and Mayr (111); they show a distal array of microvilli and ciliary structures are lacking (45,53). The Joseph cells in the dorsal wall of the cerebral vesicle may also exhibit a photoreceptive function, although these cells are not associated with pigment cells. Joseph cells have an array of microvilli on one side and occasionally a striated root has been observed (45,130).

Apart from these, there are lamellated cells in the roof of the cerebral vesicle near the Joseph cells that may have a photoreceptive function. They bear lamellar appendages that may arise from both sides of a modified cilium with 9+0 axoneme, basal body and striated rootlet (41,45,53). In another interpretation, the lamellae are derived from the cell membrane without connection to the cilia, the latter

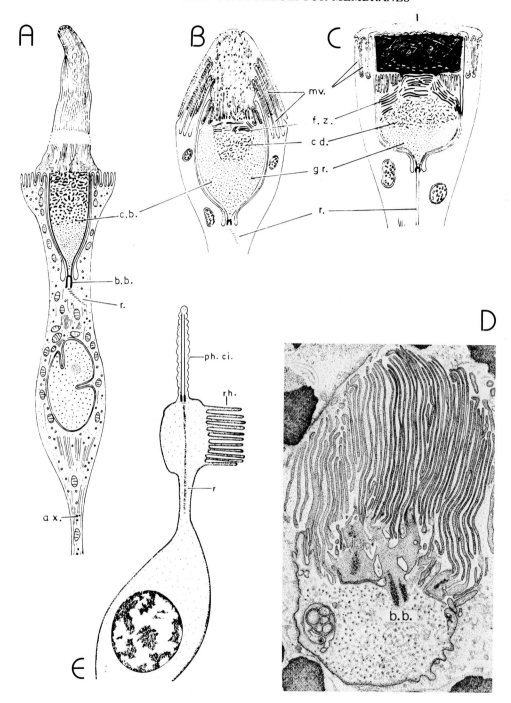

FIG. 6. A-C:Chaetognath photoreceptoral cells. A:*Sagitta*. After Eakin and Westfall (56). B:*Serratosagitta*. C:*Eukrohnia*. B and C after Ducret (37). D:Sensory process of visual cells of ascidiacean tadpole larvae. After Barnes (4). E:Sensory cell of a tornarian tadpole. After Brandenburger et al. (15).

having 9+2 or more irregular axonemes (100). Thus the presence of ciliary photoreceptors in Cephalochordata remains to be proven.

Vertebrata

Vertebrates possess one pair of lateral eyes and originally two everse median eyes. The fine structure of the light sensitive organelles in the sense cells of the lateral and median eyes is very similar and originates by infoldings of the ciliary (9+0 axonemes) membranes to form the well-known disks in rods and cones.

PHOTORECEPTOR STRUCTURE AND EVOLUTION

In discussions dealing with possible lines of photoreceptor evolution it seems worthwhile to consider what the terms "photoreceptor structure" and "monophyletic/polyphyletic origin of photoreceptors" may mean.

Most authors use the term "photoreceptoral organ", or briefly "photoreceptor" to designate rather simple light sensitive organs. Ocelli are simple photoreceptive organs that contain shading pigment either in the photoreceptive cells or in accompanying cells. More complex organs are usually called eyes, but there is no strict definition of the latter to restrict its use. The light sensitive unit of each photoreceptive cell is the photoreceptive organelle; however, the latter organelle is also frequently called "photoreceptor". The same word may thus refer to both an entire organ or just a cell organelle.

What then does "independent origin of photoreceptor structure" mean? Did the photoreceptive organelles (ciliary as well as rhabdomeric) originate several times, or was it perhaps the assembly of more or less similar organs that was achieved several times in unrelated groups by using both ancestral and new potentials? Eakin (41,43,45,46,47) asserts that there are two main lines of evolution of photoreceptive organelles, whereas Salvini-Plawen and Mayr (111) claim that there may be many lines of independent evolution of photoreceptive organs.

In addition to the problem of what structure (organ, organelle) is being considered, there is the problem as to what extent the structure may be considered as being newly evolved. Thus when asserting that a structure has originated several times, it may be assumed that the basic genetic information responsible for that structure has been created only once and, although never lost in the course of evolution, reached novel expression in supposedly unrelated groups. Or it might be assumed that this basic information has been indeed lost, yet originated as many times as the character reappeared. The latter assumption seems very unrealistic in terms of molecular genetics. Genetic information of novel characters is neither likely to be created de novo many times, nor to be lost rapidly. It need not be expressed phenotypically to be retained in the cell's genome. Evolutionarily speaking, genes are being constantly modified and combined with other genes to design altered characters and possibly new functions.

This and the morphological evidence I shall present led us (122) to speculate that there is a close structural relationship between both types (ciliary, rhabdomeric) of photoreceptor organelles and that the difference between the photoreceptor types is more of a quantitative than of a qualitative order. We have suggested that in both types the elaboration of the photoreceptor organelle is induced by a ciliary formation that may become more or less abortive (rhabdomeric type) or may develop further into a ciliary organelle (ciliary type). In other words, we interpret rhabdomeric receptors to be a modified form of their ciliary ancestors retaining most if not all of the original information required for the elaboration of a functional light sensitive organelle.

Such a modification conceivably evolved early in the evolution of the Metazoa and was adopted by the ancestral groups that gave rise to protostomous metazoans. A similar modification may have occurred a few more times among deuterostomous groups (Echinodermata, Thaliaceae, Cephalochordata) that originally retained the ciliary type of photoreceptive organelle. Both ciliary and rhabdomeric photoreceptive organelles then differentiated and specialized further, culminating in various elaborate organelles such as observed in the rods and cones of vertebrate eyes and the rhabdoms of cephalopod and arthropod eyes. Because the presence of genetic information to form the original ciliary type has been retained for initiating elaboration of otherwise rhabdomeric organelles, it is not surprising that various rudimentary ciliary structures may persist in rhabdomeric organelles or that the ciliary type may arise rather frequently in rhabdomeric groups.

Because of our hypothesis that the elaboration of photoreceptor organelles--whatever their type--is induced by a ciliary formation, various authors ascribed to us a monophyletic concept of photoreceptor evolution. This is correct if it means that the very basic genetic information for the elaboration of photoreceptor organelles originated only once, possibly in some ancestral protist. In this sense there is a monophyletic origin of photoreceptor organelles; however, most characters originate only once so this qualification is of no use for studying phylogenetic relationships. Consequently, there is no contradiction to the assertion that the rhabdomeric modification of the ancestral ciliary type has evolved independently a few times in a predominantly ciliary line and that the ciliary type has reappeared several times in a predominantly rhabdomeric line. In this sense both ciliary and rhabdomeric types have been realized independently several times, although I agree with Eakin (41,43,45,46,47) that two main trends of photoreceptor evolution may be distinguished.

Salvini-Plawen and Mayr (122) reject this view; instead they distinguish at least 40 phyletic lines. First, they consider the entire photoreceptive organ as already mentioned. Obviously, this is much more complex than tracing lines of organelle evolution. It is quite clear, for example, that branchial and pygidial ocelli of annelids and pallial and tegmental eyes in mollusks are newly acquired formations that are not homologous and thus cannot be compared directly with the cerebral ocelli of these same animals. Application of this distinction to the single components of the photoreceptoral organs, e.g., the type of receptive organelle involved is meaningless. In other words, newly evolved eyes need not exhibit newly originated receptive organelles, instead the ancestral type of organelle may well be incorporated into the secondarily acquired formation. Everybody will probably agree that both ancestral and secondarily evolved eyes are most likely to be equipped with ancestral light sensitive molecules.

In conclusion, there is no reason to believe that there are as many lines of independent evolution of photoreceptive organelles as there possibly are independent origins of photoreceptoral organs.

Secondly, recent findings suggest that the distinction between rhab-

domeric and diverticular ganglionic receptor cells is not justified.
Indeed, the previously presumed "epidermal" anterior ocelli of rotifers
are in fact extensions of cerebral neurons (25). Some of these are
exclusively ciliary (Trichocerca and Philodina, Fig. 4A and B respectively) and there is good evidence that other cerebral neurons may bear
photoreceptive cilia in rotifers, e.g., in Philodina (Fig. 3C-E). Photoreceptor cells embedded within the brain and possessing ciliary formations have been found in many other taxa and at least in some of them
are likely to be of ganglionic origin (Platyhelminthes, Gastrotricha,
Nemertini, Nematoda...).

In his most recent paper, Clement (25) advocates a polyphyletic
origin of photoreceptors. This author distinguishes three phyletic
lines of photoreceptor types in lower metazoans, all of which are
present in rotifers. Ampullae-shaped cilia, like those found in the
cerebral eyes of Philodina (Fig. 3C-E), and the anterior ocelli of
Trichocerca (Fig. 4A) are purported to represent a first line that
might be derived from the photoreceptor apparatus of phytoflagellates.
A second line would be characterized by phaosome-like photoreceptive
organelles like those described in Philodina (Fig. 4B), Platyhelminthes,
Annelida and Pogonophora. In this line there would be a progressive
evolution from ciliary to rhabdomeric types. Cylindrical and lamellar
rhabdomeres juxtaposed to a pigmented cup would represent still another
line, present in Platyhelminthes, Rotifera and other Aschelminthes,
Polychaeta and Arthropoda.

On the other hand, Clement (25) distinguishes three photosensitivi-
ties in rotifers: (1) phototaxis, which governs the orientation of the
organism, (2) photokinesis, directing their movements, and (3) photo-
periodism, which controls the production of mictic females. These photo-
sensitivities are characterized by different action spectra. Hence,
different photopigments are conceivably involved and perhaps different
photoreceptors.

I will not comment further on these hypotheses. More data are needed.
However, the idea of a possible relationship between photosensitivity,
photopigment and photoreceptor type is exciting, and the evolution of
photoreceptor structure and function may be far more complicated than we
imagine at present.

CILIARY INDUCED PHOTORECEPTOR ORGANELLES

Having treated a presumed inductive or organizing role for ciliary
structures in the elaboration of photoreceptor membranes in more general
terms, I shall now debate the morphological evidence in favor of this
idea. First, the number of exceptions to Eakin's theory has been
steadily increasing (Fig. 7). As more exceptions to his major lines in
the evolution of light-sensitive organelles arise, the more artificial
and less plausible the explanation of their independent origin appears,
unless it is assumed that both types are, in fact, closely related.

Secondly, photoreceptor cells may contain both cilia or extensions
derived from the ciliary membrane and microvilli arising from the apical
cell membrane. Such cells can only tentatively be classified as long as
it is unknown what organelles are sensitive to light; it would not be
surprising to find sense cells in which both types of organelles are
photoreceptive. For example, in the eyes of the mollusk Aplysia
receptor cells with one or a few cilia and numerous long microvilli alternate with presumed photoreceptive cells carrying equal numbers

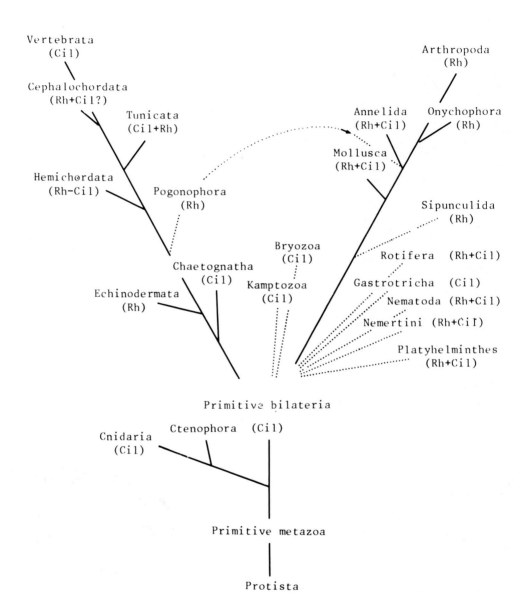

FIG. 7. Occurrence of ciliary and rhabdomeric photoreceptors within
the animal kingdom. Cil = ciliary; Rh = rhabdomeric; Rh-Cil = presumed
mixed type of receptor. After Vanfleteren and Coomans (122), modified.

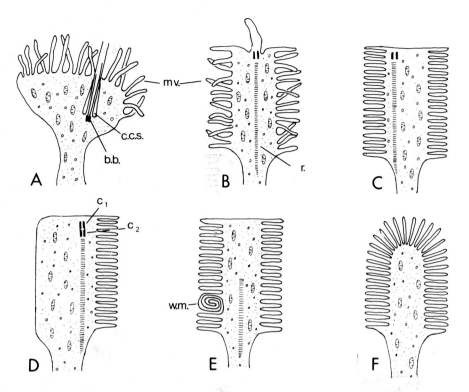

FIG. 8. Morphological sequence of rhabdomeric sensory cells with accessory ciliary structures in various stages of differentiation. A:
Phascolosoma; B:Eulalia; C:Vanadis; D:Adalia; E: Nereis; F:various
invertebrates. After Vanfleteren and Coomans (122); based on various
authors (see text).

of cilia and microvilli (Fig. 4C). Another example is that of a Tornaria
larva photoreceptor that bears a prominent cilium and an array of
microvilli (Fig. 6D). Such photoreceptors do not fit well in schemes
based on structurally unrelated organelles.

Thirdly, many rhabdomeric photoreceptors possess reduced ciliary
structures in different stages of development, where morphological
sequences of ciliary differentiation can easily be assessed (Fig. 8).
Eakin (47) interprets such ciliary formations as incidental, developmental vestiges, unrelated to the array of photosensitive microvilli.
The underlying concept is that photoreceptor cells usually differentiate
from ciliated ectodermal cells. This explanation does not account for
the observation that some of these structures, such as the rootlets may
still be very well developed, if not enlarged, in the adult stage.
There are not many, if any, structures in living cells without a
function. Furthermore, centrioles and rootlets have also been found in
developing photoreceptors of arthropods where cilia have yet to be demonstrated in embryonic cells other than sensory cells.

Centrioles lacking associated rootlets appear to be more typically
present in developing (arthropod) eyes. They might be remnants of

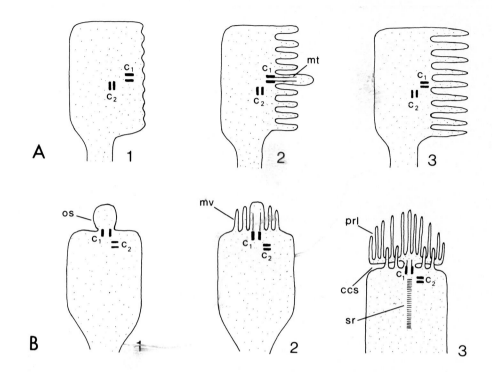

FIG. 9. Ciliary induction of the photoreceptor organelle in sensory cells of Ischryopsalis (A) and Amaroucium (B). After Vanfleteren and Coomans (122); based on Munoz-Cuevas (101) and Barnes (5).

mitotic centrioles, especially when their position is relatively removed from the rhabdomere, as in one stage of pupal eye development in Megachile (127). As pointed out by Home (77), however, rhabdomere development had already begun in that stage. In the embryonic eyes of Locusta (128), Carausius (120), Ischryopsalis (101) and, most recently, those of the honey bee Apis (62) centrioles are in close contact with the growing microvilli.

     The general picture arising from such investigations is that paired centrioles lie close to the cell membrane of the sensory cell. In a next phase they may project microtubules into a ciliary bud. Rhabdomere development begins at the time the bud is present. We interpret the inductive phase to be very transient and finished within 48 hours. Then the whole ciliary apparatus may disappear, e.g., in visual cells of Apis (Fig. 9A). The inductory process just described is very similar to the sequence of events leading to elaboration of photoreceptoral organelles of the ciliary type (Fig. 9B). It is also compatible with the organizing role of ciliary structures during ontogeny of the organ of Corti in mammals. Here developing hair cells possess a kinocilium that later may disappear, leaving only a basal body that still occupies a well defined position in relation to the stereocilia (40,63,89).

     It has been argued that, if basal bodies are organizers of photoreceptive villi, as I claim, one would expect brush borders to exhibit

evidence of ciliary structures in the embryo, if not in the adult (47).
I reject this objection. My hypothesis implies a genetically programmed
sequence of events leading to a functional light-sensitive organelle,
including membrane proliferation, production of light-sensitive
molecules and everything else needed to build a photoreceptor organelle.
Ciliary induction of membrane proliferation is just one link of a chain,
whose occurrence becomes plausible if we assume the ancestral photoreceptor organelle to be a modified cilium. This is a reasonble
assumption because cilia or flagella originally had a bimodal function
and are--apart from motility organelles--also primary sensory organelles. The primitive know-how of detecting light has been retained in
the genome during the course of evolution. Obviously, there is no
plausible reason to believe that all microvillar organelles are derived
from cilia.

    In conclusion, I believe there is reasonable evidence to suggest that
ciliary structures play an inductory role in the elaboration of both
so-called ciliary and rhabdomeric photoreceptor organelles. In fully
differentiated organelles of the ciliary type these structures are still
functional, whereas in fully developed rhabdomeric organelles ciliary
remnants may retain some function or the entire ciliary apparatus may
have disintegrated. Although this hypothesis may seem provocative, at
least it can be checked experimentally. Future eye research, especially
on the ontogeny of photoreceptors may provide more definitive answers.

ACKNOWLEDGMENTS

    I am greatly indebted to Prof. Dr. A. Coomans, who permitted me to
use freely his recent review "Phylogenetic implications of the photoreceptor structure (26). I am also indebted to Prof. Dr. Jane A.
Westfall, symposium organizer, for the kind invitation. The author
is sponsored (qualified career scientist) by the Belgian National Fund
for Scientific Research.

ABBREVIATIONS

ax., axon; b.b., basal body; b.c., basal cell; c_1,c_2, distal, proximal
centriole; c.b., conical body; c.c.s., circumciliary space; cd.,
cords; ci., cilium (cilia); cu., cuticle; e.ca., extracellular cavity;
f.z., fingerprint zone (long cords); gl.c., glial cell; gr., granules;
lam., lamellae; l.c., lens cell; mt., microtubules, mv., microvilli;
o.s., outer segment; p.c., pigment cell; p.g., pigment granule; ph.ci.,
photoreceptive cilium (cilia); pr.l., photoreceptive lamellae; r.,
rootlet; rh., rhabdomere; s.c., sensory cell; s.ci, sensory cilium;
se.c., secretory cell; s.n., sensory neurite; s.p.c., supportive cell;
s.r., striated rootlet; s.s., sensory sac; w.m., whorled membrane.

REFERENCES

1. Åkesson,B.(1958): Undersökn. Öresund, 38:1-249.

2. Barber,V.C.,Evans,E.M., and Land,M.F.(1967): Z. Zellforsch. Mikrosk.
    Anat., 76:295-312.

3. Barber,V.C., and Wright,D.E.(1969): J. Ultrastruct. Res., 26:515-528.

4. Barnes,S.N.(1971): Z. Zellforsch. Mikrosk. Anat., 117:1-16.

5. Barnes,S.N.(1974): Cell Tissue Res., 155:27-45.

6. Bedini,C.,Ferrero,E., and Lanfranchi,A.(1973): Monit. zool. ital. (N.S.), 7:51-70.

7. Bedini,C., and Lanfranchi,A.(1974): Z. Morphol. Tiere, 77:175-186.

8. Bocquet,M.(1976): J. Microsc. Biol. Cell., 25:61-66.

9. Bocquet,M.(1977): J. Ultrastruct. Res., 58:210-217.

10. Bocquet,M., and Dhainaut-Courtois,N.(1973a): J. Microsc., 18:207-230.

11. Bocquet,M., and Dhainaut-Courtois,N.(1973b): J. Microsc., 18:231-246.

12. Bouillon,J., and Nielsen,M.(1974): Arch. Biol. (Bruxelles), 85:307-328.

13. Boyle,P.R.(1969): Z. Zellforsch. Mikrosk. Anat., 102:313-332.

14. Brandenburger,J.L.(1975): J. Ultrastruct. Res., 50:216-230.

15. Brandenburger,J.L.,Woollacott,R.M., and Eakin,R.M.(1973): Z. Zellforsch. Mikrosk. Anat., 142:89-102.

16. Brooker,B.E.(1972): In:Behavioural Aspects of Parasite Transmission, edited by E.U. Canning and C.A. Wright, pp.171-180. Academic Press, London.

17. Burr,A.H., and Burr,C.(1975): J. Ultrastruct. Res., 51:1-15.

18. Burr,A.H., and Webster,J.M.(1971): J. Ultrastruct. Res., 36:621-632.

19. Burt,M.D.B., and Philips,W.L.(1969): Am. Zool., 9:622.

20. Carpenter,K.S.,Morita,M., and Best,J.B.(1974): Cell Tissue Res., 148:143-158.

21. Charles,G.(1966): In: Physiology of Mollusca., Vol. 2, edited by K. Wilbur and C.M. Yonge, pp. 455-521. Academic Press, New York and London.

22. Clark,A.W.(1967): J. Cell Sci., 2:314-348.

23. Clément,P.(1975): J. Microsc. Biol. Cell., 22:69-86.

24. Clément,P.(1977): Arch. Hydrobiol., 8:270-297.

25. Clément,P.(1980): Hydrobiologia, 73:93-117.

26. Coomans,A.(1981): In:Origine dei Grandi Phyla dei Metazoi. Accademia Nazionale dei Lincei, Roma (in press).

27. Croll,N.A.,Evans,A.A.F., and Smith,J.M.(1975): Comp. Biochem. Physiol., 51(A):139-143.

28. Croll,N.A.,Riding,I.L., and Smith,J.M.(1972): Comp. Biochem. Physiol., 42(A):999-1009.

29. Dhainaut-Courtois,N.(1965): C.R. Acad. Sci. (Paris), 261:1085-1088.

30. Dilly,P.N.(1961): Nature (London), 191:786-787.

31. Dilly,P.N.(1964): Q. J. Microsc. Sci., 105:13-20.

32. Dilly,P.N.(1969a): Z. Zellforsch. Mikrosk. Anat., 96:63-65.

33. Dilly,P.N.(1969b): Z. Zellforsch. Mikrosk. Anat., 99:420-429.

34. Dilly,P., and Wolken,J.(1973): Micron, 4:11-29.

35. Dorsett,D.A., and Hyde,R.(1968): Z. Zellforsch. Mikrosk. Anat., 85:243-255.

36. Ducret,F.(1975): Cah. Biol. Mar., 16:287-300.

37. Ducret,F.(1978): Zoomorphologie, 91:201-215.

38. Durand,J.P., and Gourbault,N.(1975): Ann. Speleol., 30:129-135.

39. Durand,J.P., and Gourbault,N.(1977): Can. J. Zool., 55:381-390.

40. Duvall,A.J.,Flock,A., and Wersall,J.(1966): J. Cell Biol., 29:497-505.

41. Eakin,R.M.(1963): In: General Physiology of Cell Specialization, edited by D. Mazia and A. Tyler, pp. 393-425, McGraw-Hill, New York.

42. Eakin,R.M.(1965): Am. Zool., 5:249.

43. Eakin,R.M.(1966a): Cold Spring Harbor Symp. Quant. Biol., 30:363-370.

44. Eakin,R.M.(1966b): 6th Int. Congr. Electron Microsc. (Kyoto) 2:507-508.

45. Eakin,R.M.(1968): In:Evolutionary Biology, Vol. 2, edited by T. Dobzhansky, M.K. Heckt and W.C. Steere, pp. 194-242. Appleton-Century-Crofts, New York.

46. Eakin,R.M.(1972): In:Handbook of Sensory Physiology, Vol. VII/1, edited by H.J.A. Dartnall, pp. 625-684. Springer-Verlag, Berlin, Heidelberg, New York.

47. Eakin,R.M.(1979): Am. Zool., 19:647-653.

48. Eakin,R.M., and Brandenburger,J.L.(1967): J. Ultrastruct. Res., 18:391-421.

49. Eakin,R.M., and Brandenburger,J.L.(1979):Zoomorphologie, 92:191-200.

50. Eakin,R.M., and   Kuda,A.(1971): Z. Zellforsch. Mikrosk. Anat.,
    112:287-312.

51. Eakin,R.M.,Brandenburger,J.L., and Barker,G.M.(1980): Zoomorphologie,
    94:225-239.

52. Eakin,R.M.,Martin,G.C., and Reed,C.T.(1977): Zoomorphologie, 88:1-18.

53. Eakin,R.M., and Westfall,J.A.(1962a): J. Ultrastruct. Res., 6:531-539.

54. Eakin,R.M., and   Westfall,J.A.(1962b): Proc. Natl. Acad. Sci. USA,
    48:826-833.

55. Eakin,R.M., and Westfall,J.A.(1964a): Z. Zellforsch. Mikrosk. Anat.,
    62:310-332.

56. Eakin,R.M., and Westfall,J.A.(1964b): J. Cell Biol., 21:115-132.

57. Eakin,R.M., and Westfall,J.A.(1965a): J. Ultrastruct. Res., 12:46-62.

58. Eakin,R.M., and Westfall,J.A.(1965b): Z. Zellforsch. Mikrosk. Anat.,
    68:278-300.

59. Eakin,R.M.,Westfall,J.A., and Dennis,M.J.(1967): J. Cell Sci.,
    2:349-358.

60. Ehlers,B., and Ehlers,U.(1977a): Zoomorphologie, 87:65-72.

61. Ehlers,B., and   Ehlers,U.(1977b): Zoomorphologie, 88:163-174.

62. Eisen,J.S., and Youssef,N.N.(1980): J. Ultrastruct. Res., 71:79-94.

63. Engström,H., and Ades,H.W.(1973): In: The Ultrastructure of Sensory
    Organs, edited by I. Friedmann, pp. 83-151. North-Holland/
    American Elsevier, Amsterdam, London, New York.

64. Fischer,A., and Brökelmann,J.(1966): Z. Zellforsch. Mikrosk. Anat.,
    71:217-244.

65. Fischer,F.P.(1979): Zoomorphologie, 92:95-106.

66. Fournier,A.(1975): Z. Parasitenkd., 46:203-209.

67. Fournier,A., and Combes,C.(1978): Zoomorphologie, 91:147-155.

68. Gorman,A.L.F.,McReynold,J.S., and Barnes,S.N.(1971): Science,
    172:1052-1054.

69. Greuet,C.(1978): Cytobiologie, 17:114-136.

70. Haas,W., and Kriesten,K.(1978): Zoomorphologie, 90:253-268.

71. Hermans,C.O.(1969): Z. Zellforsch. Mikrosk. Anat., 96:361-371.

72. Hermans,C.O., and Cloney,R.A.(1966): Z. Zellforsch. Mikrosk. Anat., 72:583-596.

73. Hermans,C.O.,and Eakin,R.M.(1969): Z. Zellforsch. Mikrosk. Anat., 100:325-339.

74. Hermans,C.O., and Eakin,R.M.(1971): Proc. Int. Symp. Biol. Sipuncula and Echiura I, Kotor, 1970:229-237.

75. Hermans,C.O., and Eakin,R.M.(1974): Z. Morphol. Tiere, 79:245-267.

76. Home,E.M.(1972): Tissue Cell, 4:227-234.

77. Home,E.M.(1975): Tissue Cell, 7:703-722.

78. Home,E.M.(1976): Tissue Cell, 8:311-333.

79. Horridge,G.A.(1964): Q. J. Microsc. Sci., 105:311-317.

80. Hughes,H.P.I.(1970): Z. Zellforsch. Mikrosk. Anat., 106:79-98.

81. Hughes,R.L.Jr., and Woollacott,R.M.(1978): Zoomorphologie, 91:225-234.

82. Isseroff,H., and Cable,R.M.(1968): Z. Zellforsch. Mikrosk. Anat., 86:511-534.

83. Juberthie,C., and Munoz-Cuevas,A.(1973): C.R. Acad. Sci. (Paris), 276:2537-2539.

84. Kearn,C.G., and Baker, N.O.(1973): Z. Parasitenkd.,41:239-254.

85. Kerneis,A.(1968a): J. Microsc., 7:40a.

86. Kerneis,A.(1968b): Z. Zellforsch. Mikrosk. Anat., 86:280-292.

87. Kerneis,A.(1975): C.R. Acad. Sci., (Paris), 273:372-375.

88. Kerneis,A.(1975): J. Ultrastruct. Res., 53:164-179.

89. Kikuchi,K.,and Hilding,D.(1965): Acta Otolaryngol. (Stockh), 60:207-222.

90. Kishida,Y.(1967): Sci. Rep. Kanazawa Univ., 12:75-110.

91. Krisch,B.(1973): Z. Zellforsch. Mikrosk. Anat., 142:241-262.

92. Krasne,F.B., and Lawrence,P.A.(1966): J. Cell Sci., 1:239-248.

93. Kuster,J.E.(1980): Cell Tissue Res., 206:123-138.

94. Lanfranchi,A.,Bedini,C., and Ferrero,E.(1981): 3rd International Symposium of Turbellaria. (in press).

95. Levi,P.,and Levi,C.(1971): J. Microsc., 11:425-432.

96. Lyons,K.M.(1972): In:Behavioural Aspects of Parasite Transmission, edited by E.U. Canning and C.A. Wright, pp. 181-199. Academic Press, London.

97. MacRae,E.K.(1964): J. Ultrastruct. Res., 10:334-349.

98. MacRae,E.K.(1966): Z. Zellforsch. Mikrosk. Anat., 75:469-484.

99. Merker,G.,and Vaupel von Harnack,M.(1967): Z. Zellforsch. Mikrosk. Anat., 81:221-238.

100. Meves,A.(1973): Z. Zellforsch. Mikrosk. Anat., 139:511-532.

101. Munoz-Cuevas,A.(1975): C.R. Acad. Sci. (Paris), 280:725-727.

102. Nørrevang,A.(1974): Zool. Anz., 193:297-304.

103. Penn,P.E.,and Alexander,C.G.(1980): Mar. Biol., 58:251-256.

104. Pond,G.,and Cable,R.M.(1966): J. Parasitol., 52:483-493.

105. Röhlich,P.(1966): Z. Zellforsch. Mikrosk. Anat., 73:165-173.

106. Röhlich,P.,Aros,B.,and Viragh,S.(1970): Z. Zellforsch. Mikrosk. Anat., 104:345-357.

107. Röhlich,P.,and Török,L.J.(1963): Z. Zellforsch. Mikrosk. Anat. 60:348-368.

108. Rosen,M.D.,Stasek,C.R., and Hermans,C.O.(1978): The Veliger, 21:10-18.

109. Rosen,R.M.,Stasek,C.R., and Hermans,C.O.(1979): The Veliger, 22:173-178.

110. Ruppert,E.E.(1978): In:Settlement and Metamorphosis of Marine Invertebrate Larvae, edited by F.S. Chia and M.E. Rice, pp. 65-81. Elsevier/North-Holland Biomedical Press, New York.

111. Salvini-Plawen,L.v.,and Mayr,E.(1977): In: Evolutionary Biology, Vol. 10, edited by M.K. Hecht,W.C. Steere and B. Wallace, pp. 207-263. Plenum, New York.

112. Schwalbach,G.,Lickfeld,K.G.,and Hahn,M.(1963): Protoplasma, 56:242-273.

113. Short,R.B.,and Gagné,H.T.(1975): J. Parasitol., 61:69-74.

114. Siddiqui,I.A.,and Viglierchio,D.R.(1970a): J. Nematol., 2:274-276.

115. Siddiqui,I.A.,and Viglierchio,D.R.(1970b): J. Ultrastruct. Res., 32:558-571.

116. Singla,C.L.(1974): Cell Tissue Res., 149:413-429.

117. Singla,C.L.(1975): J. Ultrastruct. Res., 52:333-339.

118. Stewart,A.(1966): Am. Zool., 6:615-616.

119. Storch,V., and Mortiz,K.(1971): Z. Zellforsch. Mikrosk. Anat., 117:212-225.

120. Such,J.(1969): C.R. Acad. Sci. (Paris), 268:948-949.

121. Teuchert,G.(1976): Zoomorphologie, 83:193-207.

122. Vanfleteren,J.R., and Coomans,A.(1976): Z. Zool. Syst. Evolutions-forsch., 14:157-169.

123. Vaupel von Harnack,M.(1963): Z. Zellforsch. Mikrosk. Anat., 60:432-451.

124. Verger-Bocquet,M.(1977): Biol. Cellulaire, 30:65-72.

125. Vernet,G.(1970): Z. Zellforsch. Mikrosk. Anat., 104:494-506.

126. Vernet,G.(1974): Ann. Sci. Nat. Zool., 16:27-36.

127. Wachmann,E., and Hennig,A.(1974): Z. Morphol. Tiere, 77:337-344.

128. Wachmann,E.,Richter,S., and Schricker,B.(1973): Z. Morphol. Tiere, 76:109-128.

129. Wells,M.J.(1966): In:Physiology of Mollusca, Vol. 2, edited by K.M. Wilbur and C.M. Yonge, pp. 523-545. Academic Press, New York and London.

130. Welsch,U.(1968): Z. Zellforsch. Mikrosk. Anat., 86:252-261.

131. White,R., and Walther,J.(1969): Z. Zellforsch. Mikrosk. Anat., 95:102-108.

132. Whittle,A.C., and Golding,D.W.(1974): Cell Tissue Research, 154:379-398.

133. Wolken,J.J.(1977): J. Protozool., 24:518-522.

134. Woollacott,R.(1974): Dev. Biol., 40:186-195.

135. Woollacott,R., and Eakin,R.M.(1973): J. Ultrastruct. Res., 43:412-435.

136. Woollacott,R.H., and Zimmer,R.L.(1972): Z. Zellforsch. Mikrosk. Anat., 123:458-469.

137. Yamamoto,T.,Tasaki,K.,Sugawara,Y., and Tonosaki,A.(1965): J. Cell Biol., 25:345-359.

138. Yamamoto,M., and Yoshida,M.(1978): Zoomorphologie, 90:1-17.

139. Yanase,T., and Sakamoto,S.(1965): Zool. Mag., 74:238-242.

140. Young,J.Z.(1962): Philos. Trans. R. Soc. Lond. (Biol), 245:1-18.

141. Young,J.A.(1971): The Anatomy of the Nervous System of Octopus vulgaris. Clarendon Press, Oxford.

142. Zahid,Z.R., and Golding,D.W.(1974): Cell Tissue Res., 149:567-576.

Visual Cells in Evolution, edited by Jane A. Westfall,
Raven Press, New York © 1982.

On the Polyphyletic Origin of Photoreceptors

Luitfried v. Salvini-Plawen

Zoologisches Institut, Universität Wien, Wien 1, Austria

Thanks to the extensive studies by Richard M. Eakin and his research
group on the structure of photoreceptor cells we recognize that their
major specialization is the provision of large surface areas in a relatively superficial position upon which the light impinges. Those
fundamental investigations also demonstrated the presence of enlarged
membranes of two different kinds: the modified plasma membrane of a
cilium or flagellum (ciliary type), or the elaboration of the distal
cell membrane proper without relation to cilia, (rhabdomeric type sensu
Eakin). The occurrence of the two types among the animal groups investigated led Eakin (15, 16, 17) to develop a phylogenetic concept of two
evolutionary lines including the Coelenterates and Deuterostomia on the
one hand (ciliary line), and the flatworms and Protostomia on the other
(Eakin's rhabdomeric line). Although subsequent findings increasingly
weakened that concept (6, 30) and contradicted the phylogenetic value of
the two types, it appears to be generally accepted (e.g., 60) and is
even used as essential support for phylogenetic reconstruction (e.g.,
45).

Vanfleteren and Coomans (62) as well as Salvini-Plawen and Mayr (55)
presented evidence that refutes the concept that evolution of photoreceptors occurs along two basic phylogenetic lines. Evidence presented
by both groups coincides in two essential features: 1) in the
rhabdomeric line, the presence "of ciliary type of photoreceptors...
appears to be rather commonly associated with secondary formations"
(62, p.164); an equivalent statement by Salvini-Plawen and Mayr
(55, p.255) is that "most photoreceptors with the ciliary type of
structure can be shown to represent phylogenetically young acquisitions"
and 2) it is concluded that the photoreceptor type is not conservative
enough to assign the phyla to main lines of evolution, but it may be
useful "in studying closer phylogenetic relationships" (62, p.165).
Salvini-Plawen and Mayr (55, p.255) also concluded "that selective
pressures for photoreceptors did not become strong enough to be re-
sponded to until the radiation into major lines (phyla, classes) of
recent organisms was well established."

In view of the diversity of photoreceptor differentiation of unequiva
lent phylogenetic value and of different organogenetic appearance Eakin
(18) updates his concept with a modified argument. Let us consider
Eakin's new position and discuss the concepts of Vanfleteren and Coomans
(62) prior to offering some additional arguments for a polyphyletic
origin of photoreceptors, especially with respect to the cerebral organs

within the Protostomia.

1. A MUTATION THEORY

Salvini-Plawen and Mayr's main criticism of Eakin's concept is a
phylogenetic one. According to rigorous homologic comparisons (as far
as present knowledge allows), the (ultra-)structurally conceivable
differentiation of photoreceptors did not evolve along two major
evolutionary pathways, but rather along many lines at minor levels of
supraspecific grades. An essential conclusion is that the ultrastructural characteristics tend to demonstrate homology only within
limited taxa. Eakin (18, p.650) accepts the presence of ciliary
photoreceptors in protostomes as "indeed cenogenetic, i.e., secondarily
evolved, structures." He also holds "that integumentary and cerebral
ocelli are nonhomologous"; the "best explanation for these exceptions
is probably Salvini-Plawen and Mayr's theory of independent evolution"
(18, p.651). Thus, both theories appear to coincide, fully supporting
a polyphyletic differentiation, since the occurrence of the rhabdomeric
types within Eakin's ciliary line of evolution of photoreceptors "may
be other instances of cenogenetic evolution" (18, p.651).

Confronted with such a situation, Eakin (18, p.652) modifies his
concept as follows: All ciliary membranes are able "to build and to use
the chief structural protein of the ciliary membrane--an opsin conju-
gated with vitamin A--to capture photons", and that basic "know-how"
continues through the coelenterates and deuterostomes (ciliary line) as
well as conservatively down to the exceptions" in his rhabdomeric line.
Somewhere above the coelenterates,however, mutations should have endowed the sensory cell plasmalemma with photosensitivity, thus finally
becoming rhabdomeric photoreceptors and representing a commonly adapted
(synapomorphous) character in the respective groups. Unfortunately,
there are exceptions in this respect as well, so "that photosensitive
microvilli arose independently and secondarily in the echinoderms"
(18, p.652), as well as in Thaliacea (Tunicata), and in Cephalochordata,
and in Tornariae, etc. Why not likewise in Mollusca or Annelida?

What remains of Eakin's concept now? There is a diffuse photosensitivity in ectodermal ciliary cells owing to their property of in-
corporating photopigment into the ciliary membrane (see dermal light
sense; 37, 42), and their polyphyletic differentiation towards photoreceptors (see 55). Some of those lines of differentiation adapted to
a more particular elaboration by shifting the light-gathering power to
the cell membrane, and only the line of Spiralia is accepted as representing the actual rhabdomeric line (including lines 17, 29, 31, and 34
in 55, p.218). Even without further discussion of such representation,
there are about 20 ultrastructurally confirmed lines of convergent photoreceptor evolution that stand in full accord with Salvini-Plawen and
Mayr's (55) principal conclusions (Fig. 1).

2. AN INDUCTION THEORY

Vanfleteren and Coomans (62) agree with Salvini-Plawen and Mayr on
two essential points: that the ciliary type of photoreceptor in
protostomes is a secondary acquisition, and that the phylogenetic value
of photoreceptor structure is limited to lower taxa of supraspecific
units. Nevertheless, Vanfleteren and Coomans take the viewpoint of a
pseudo-monophyletic evolution of photoreceptors, postulating that all

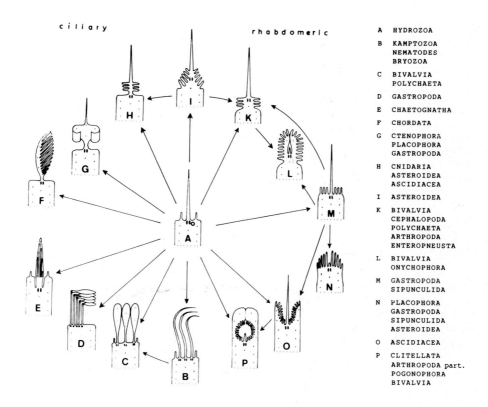

ciliary **rhabdomeric**

A	HYDROZOA
B	KAMPTOZOA NEMATODES BRYOZOA
C	BIVALVIA POLYCHAETA
D	GASTROPODA
E	CHAETOGNATHA
F	CHORDATA
G	CTENOPHORA PLACOPHORA GASTROPODA
H	CNIDARIA ASTEROIDEA ASCIDIACEA
I	ASTEROIDEA
K	BIVALVIA CEPHALOPODA POLYCHAETA ARTHROPODA ENTEROPNEUSTA
L	BIVALVIA ONYCHOPHORA
M	GASTROPODA SIPUNCULIDA
N	PLACOPHORA GASTROPODA SIPUNCULIDA ASTEROIDEA
O	ASCIDIACEA
P	CLITELLATA ARTHROPODA part. POGONOPHORA BIVALVIA

FIG. 1. Polyphyletic radiation of photoreceptors derived from epidermal cells with unmodified/mixed type of ultrastructure (cilium plus micro-villi) exhibiting general photosensitivity (dermal light sense; see Fig. 2, I and Ia). Slightly modified from Salvini-Plawen and Mayr (55).

photoreceptive plasma membranes are induced by ciliary structures. This assumption that the distal centriole (basal body, kinetosome, axial centriole) or other ciliary structure should be the organizer not only for the elaboration of membranes of the cilia provided with photo-sensitive pigment, but also for purely microvillar differentiation of rhabdomeric photoreceptors has been disputed by Eakin (18, 19). More-over, the transitory appearance of ciliary structures (centrioles, roots, etc.) in rhabdomeric photoreceptors may be considered as morpho-genetic remnants through recapitulation (biogenetic rule), including a later prominent microvillus as a vestige of the cilium itself.

There are, however, two other arguments: 1. An induction theory does not explain, <u>why</u> in several instances a rhabdomeric (microvillar) elaboration is induced rather than an expected ciliary type (as in <u>Amaroucium</u>, 62). Since the inductor itself would be an organizer normally associated with the cilium, a shift of photoreceptive power to the cell membrane cannot be effected by the ciliary inductor <u>per se</u> as

long as cilia are present (compare Littorina, Aplysia, Phascolosma,
Tornariae, echinoderms, etc., 8, 55). Any other change, however, be it
endogenous (mutation, 18) or exogenous (response to a different function
62), dissociates that organizer from its inductor function. 2. Secondly,
the induction theory of Vanfleteren and Coomans expresses a polyphyletic
differentiation of photoreceptor organs. Since homology is best defined
as coincidence of properties effected by identical information through
inheritance (56), advanced induction would be a homologous property of
all ciliary cells with diffuse photosensitivity. The derivatives,
namely, dissimilar elaborations of distinct photoreceptors in various
lines at a different time are the best homoiologous organs, i.e., analogous organs based upon that common ancestral property (56). The
differentiation and establishment of photoreceptor organs within specific evolutionary lines, is thus a polyphyletic reaction to respective
selective pressures of an already existent diffuse photosensitivity,
perhaps associated with a "recallable" organizer. The building material
(cells) may be homologous (monophyletic); the organelles produced by
them (photoreceptors), however, are polyphyletic.

3. DIFFUSE PHOTOSENSITIVITY AND FUNCTION

The concepts of mutation theory and induction theory are not satis-
factory. There is no reasonably elucidative explanation as to why on
the one hand the ciliary membrane is elaborated to become a photoreceptive organelle, and in other instances the cell membrane proper is
so elaborated. A multiple parallel mutation appears to be unsupported
(although conceivably possible), and a postulated overall inductor of
ciliary structure cannot be accepted (see above). Salvini-Plawen and
Mayr (55, p.247 and 254-5) advanced the hypotheses that (1) primitive,
nonspecialized cells exhibiting a general property for photosensitivity
(diffuse dermal light sense) were provided both "with simple cilia and
some microvilli" only later to become "canalized by selection into one
of the specialized types" of elaborate photoreceptors, and (2) "the
structural organization of photoreceptors in a phyletic line can change
in the course of evolution according to new requirements." Accordingly,
a strong functional dependence of structural differentiation must be
implied, but there is little information from a physiological approach;
hence, there is still poor understanding of functional correlation.
Nevertheless, detection of changes in light intensity (decrease: light/
shadow reaction) with a hyperpolarizing "off-response" of receptor potential appears to be correlated with the ciliary type of structure in
very slow moving or sessile animals (55, p.249 and 255; for "exceptions"
see below: Salpa). There is as yet no strict correlation between microvillar photoreceptors (existence and/or increase of intensity of light)
and the direction of incident light (orientation) and analysis of the
environment (image formation). Accordingly, it may be premature to try
to discuss the question of how an ultrastructural type of elaboration
corresponds to a certain functional requirement.

To analyze additional dependences Salvini-Plawen and Mayr (55) have
argued that differentiation of photoreceptors reflects adaptation to
functional requirements based upon unequal prerequisites (Fig. 2):
diffuse light sensitive cells, prior to the elaboration of photoreceptors
(I) were originally characterized by cilia as well as by a few microvilli (see also 13 and 65). Later differentiation with respect to
functional requirements resulted in fairly conservative, not yet modified,

u l t r a s t r u c t u r e

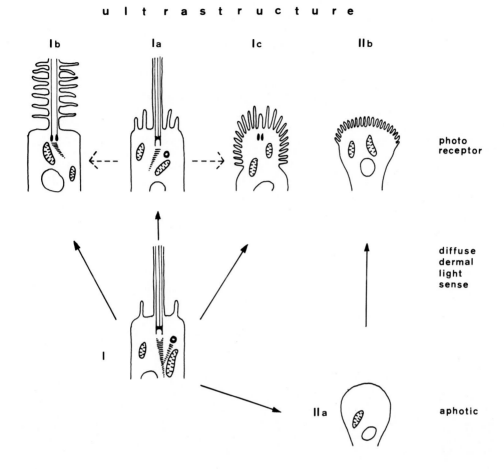

FIG. 2. Scheme of presumed differentiation of the types of photoreceptor structure. I, monociliate ectodermal cell with microvilli as a
probable stem cell (see Fig. 1); Ia, mixed type of photoreceptor; Ib,
ciliary type of photoreceptor; Ic, rhabdomeric type of photoreceptor.
IIa, aciliate and unpleated, neural/ganglionic cell (subdermal); IIb,
diverticular type of photoreceptor.

mixed types of structure with unpleated cilia and/or microvilli (Ia; see
Anthomedusae–Leuckartiara, Sipunculida–Phascolosoma, echinoderms, etc.),
or in ciliary types with differently modified cilia lacking microvilli
(Ib), or in characteristically rhabdomeric types of microvillar structure
with vestigial cilia (Ic). Alternatively, cells provided with diffuse
photosensitivity underwent regression of their cilia (as neurons and
ganglion cells) prior to eventual elaboration of a distinct photoreceptor (IIa,b). This resulted in ganglionic or diverticular types of
microvillar structure (compare Branchiostoma, Salpa, and other; see 3,
pp.8–10; 34, p.175; 55, p.226; 65, pp.611 and 615 ff; also 44, p.287).

Eakin's criticism (18) of the ganglionic diverticular type (II) is
justified to only a limited extent (without mentioning that his examples
do not belong to this type); one can only retrospectively typify a microvillar photoreceptor as belonging either to the diverticular (II) or to
the rhabdomeric type (Ic) with totally reduced ciliary structures (which
can easily include misinterpretations). Yet Branchiostoma appears to be
a typical ganglionic type and the photoreceptor of Salpa gives a hyperpolarizing "off-response" (light/shadow reaction) as do other sessile
animals with only a ciliary type photoreceptor (see 25). This may demonstrate that the requisite structure for hyperpolarizing receptor potentials, namely, a cilium, was not "available" when the cells adapted to
their functional requirement (see also 65). Accordingly, an aciliate
and unpleated neural/ganglionic photosensitive cell (3, pp.8 and 63)
could only differentiate as a diverticular structure. Such a "misused"
structure consequently "seems to undergo much less evolutionary change"
(55, p.247). In the case of more specific requirements it appears to be
supplemented by other photosensitive cells (such as the distal cell in
Nerilla, see 19) or to be superseded by new photoreceptors (see
Sipunculida or Polychaeta below).

4. POLYPHYLETIC MICROVILLAR PHOTORECEPTORS IN SPIRALIA

Eakin's rhabdomeric line of evolution has recently been restricted to
include only the cerebral organs of protostomes (18, 19). Even though
not all representatives have been investigated ultrastructurally, such
photoreceptors are present in Platyhelminthes, Nemertini, Kamptozoa,
Rotatoria, Gastrotricha, Nematoda, Kinorhyncha (Echinoderes), Mollusca,
Sipunculida, Polychaeta, and Arthropoda s.l. Are they all homologous
organs?
What criterion led Eakin (18) to conclude that the echinoderms possess secondarily evolved rhabdomeric photoreceptors? Nothing other
than a general knowledge of structural morphology, comparative anatomy,
or embryology which contradicts any closer phylogenetic affinity between
Echinodermata and Spiralia. If, consequently, the ultrastructural result
does not reflect actual relationship in that case, why does Eakin rely
on it in protostome groups? Simply because other information indicates
to us a fairly close relationship between these groups. Accordingly,
ultrastructural configuration of photoreceptors can only serve as added
support to an already well-established relationship based upon other
criteria. This includes the fact that the relationship of two groups
at higher grades of supraspecific taxa must first be examined and
established before photoreceptors can additionally serve to support their
close connection; otherwise, they only can serve for closer relationships
at the lower level of supraspecific organization.
In consideration of these reflections, an accurate analysis of the
conditions of homology as regards cerebral photoreceptors in Spiralia
will be presented. The commonly-adapted (synapomorphous) organs must be
traced out and confirmed by other data relevant to the homology theorem,
and "the possibility that two unrelated evolutionary lines have evolved
similar structures to perform the same function must be carefully considered" owing to the "limitation of organic design" (46, p.659).

Kamptozoa and Gastrotricha

In strong opposition to Eakin's concept, Salvini-Plawen and Mayr attributed the microvillar cerebral photoreceptors in question to five independent evolutionary lines (nos. 13, 17, 29, 31, 34 in 55, p.218).

The Gastrotricha must now be eliminated as an additional line since it possesses organs of a typical ciliary structure (see 61) as does the Kamptozoa line. They represent the primary two rebutting pieces of evidence to Eakin's postulation.

Turbellaria and Nemertini

A seeming discrepancy arises within the photoreceptors in Turbellaria and Nemertini assigned by Salvini-Plawen and Mayr to two different lines (nos. 13 and 34). The ocelli in Turbellaria have very different arrangements and relations to the sensory system (see 34). In most groups however, one to three pairs of ocelli are present near to or come directly from the brain (34), to which the eucerebral ocelli of the Acoela may or may not belong (see 65). With respect to their ultrastructural coincidence (only a distal brush of microvilli without cilia or their rudiments) a hereditary transmission in the evolutionary lines from lower flatworms via Dalyelloidea to Trematoda-Monogenea and miracidia-cercariae (Digenea) respectively can be accepted. On the other hand, Polycladida, several Tricladida, and some other Turbellaria possess numerous ocelli that vary broadly in their position, number, and origin, thus demonstrating that our understanding of their establishment is still somewhat weak; further, the arrangement of ocelli in juvenile and adult animals may also vary considerably (38). Special interest must be given to the polyclads in which the numerous photoreceptors are arranged in three more or less outlined groups, namely, pericerebral ocelli, tentacular ocelli, and marginal ocelli. Although all three groups are pigment-cup ocelli beneath the epidermis, only the tentacular and marginal ocelli can ontogenetically be equated as clearly epidermal eyes (see 38). No homology can be argued for the (peri-) cerebral photoreceptors. In cotylean polyclads, one pair of additional photoreceptors is present in immediate contact with the anteriodorsal section of the brain (eucerebral ocelli) and develops in situ after the late larval stage (see 38). It may be presumed that these ocelli are rudimentary homologues to the 1-3 pairs of organs in other turbellarian groups (see above).

Ultrastructural investigations on Lobophora larvae (Müller's and Götte's larvae) of estuarine Polycladida (49) revealed that the single or paired epidermal ocelli associated with the sensory apical organs possess a pigmented receptor cell with cilia modified to form flattened membranes. On the other hand, paired or single (peri)cerebral ocelli embedded within the cerebral cells of the larva consist of several multi-ciliated cells as well as several microvillar cells (in one case with a ciliary root); this demonstrates mixed photoreceptors. Owing to a lack of ATP-ase arms on the outer doublets of ciliary microtubules they likewise may be assigned to the ciliary type (compare Kamptozoa; see 55). The only adult polyclad so far ultrastructurally investigated with respect to their photoreceptors is Notoplana acticola (40). Both the pericerebral ocelli (40, Fig. 5) as well as the epidermal (tentacular) photoreceptors (40, but without documentation) are of purely microvillar structure. Although both Stylochus (Götte's larva) and Notoplana possessing similar larvae) belong to the Acotylea and their pericerebral and epidermal ocelli are homologous (for restriction see below), there is no correlation between the structures of the respective photoreceptors.

In Nemertini, the organization of the Pilidium larvae (without photoreceptors) suggests a closer relationship to the Turbellaria-Polycladida
(see 53), with which other characters also coincide. Except for a few
species, most nemertines have several or numerous pigment-cup ocelli
beneath the epidermis anterior but never in close contact with the
brain. They correspond to the epidermal and/or pericerebral ocelli in
polyclads also in dispersion and variability of number. Indeed, there is
great probability of a direct mutual homology as supported by the
epidermal origin (rhabdomeric types in Lineus with rudimentary cilium)
and the addition/subdivision of ocelli during development in both Polycladida and Nemertini.

Summing up the present state, both lines advanced by Salvini-Plawen
and Mayr (55), namely, line 13 (24) including epidermal (tentacular)
ocelli in larval and adult Turbellaria and line 34 (54) including the
common, paired eu-cerebral ocelli are well established. In addition,
line 13 (25) including the (peri-)cerebral ocelli of polyclads and
probably also the Nemertini line 13 (42) stand on developmental and
structural reasons as an independent evolutionary pathway. Both lines
13 (24) and 13 (25), however, belong to the group with ciliary photoreceptors and hence constitute another bit of evidence rebutting Eakin's
concept.

There still remains the discrepancy of ocellar structure in larval
and adult Acotylea (Stylochus versus Notoplana). The epidermal ocelli
of the larvae with purely ciliary structures "were found only in animals
collected in the Indian River estuary" but not in oceanic larvae
(49, p.80); accordingly, it is only with reservation that they can be
homologized with the epidermal (tentacular) photoreceptors in Notoplana
"collected in tidal pools" (40, p.470). Their mutual continuance can
only be retained by replacement of the purely ciliary receptor cells by
new ones that become microvillar, because in the larval receptors no
microvilli appear to be present for subsequent elaboration. Such
substitution, although taking place in the opposite direction, possibly
also occurred in some Gastropoda such as Pterotrachea (Heteropoda) when
compared with the mixed type of photoreceptors in Aplysia and the
rhabdomeric receptors of other snails (see 32).

On the other hand, for the pericerebral ocelli in Turbellaria-
Polycladida we may propose that after metamorphosis a change in requirement occurs that results in a unique elaboration of rhabdomeric cells
in the mixed organ (with abandonment of the ciliary receptor cells). A
similar condition, observed in the photoreceptors at the mantle edge in
Bivalvia (Fig. 3), may support a later evolutionary separation of the
two types of receptors in mixed organs (see 55): the everse=converse
lens eyes of Tridacna show a mixed ultrastructural appearance with some
rhabdomeric but with a predominance of ciliary receptor cells. The
homologous, inverted retina of Cardium is solely made up of ciliary type
receptors. The proximal retina in eyes of Pecten no doubt also homologous to the inverted retina of Cardium is made up of rhabdomeric type
cells; it is supplemented by ciliary cells of the distal retina, which
correspond to the inverted ciliary sense organ of Cardiacea (Tridacna,
Cardium, see 4, 5, 35, 55, 59).

Sipunculida and Polychaeta

Salvini-Plawen and Mayr (55) considered the inverse pigment-cup
ocelli in Sipunculida and Polychaeta homologous with the (eu-)cerebral

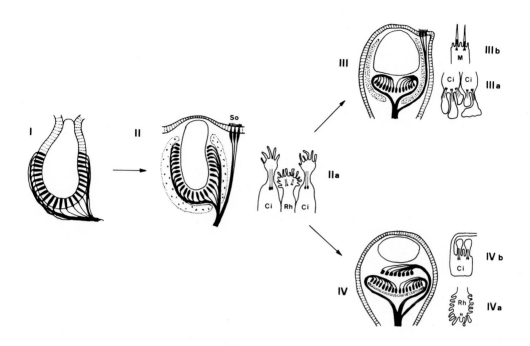

FIG. 3. Morphological sequence of differentiation in eyes at the mantle
edge of Pectenoidea (I and IV) and Cardioidea (II and III), Mollusca,
Bivalvia. I, converse pinhole eye of Lima squamosa. II, converse closed
lens eye and ciliary sense organ of Tridacna maxima; IIa, ultrastructure of the retina with ciliary and rhabdomeric type of receptors.
III, lens eye with inverted retina and ciliary sense organ of Cardium
(C. muticum); IIIa, ultrastructure of retinal cells (C. edule:ciliary
type); IIIb, ultrastructure of sense organ (C. edule:mixed type). IV,
lens eye of Pecten maximus with inverted proximal retina as in Cardium
and with an inverted ciliary sense organ representing the distal retina;
IVa, ultrastructure of the proximal retina (rhabdomeric type), IVb,
ultrastructure of the distal retina (ciliary type). Ci, ciliary type,
M, mixed type, Rh, rhabdomeric type, So, ciliary sense organ. Combined
after Barber et al. (4), Barber and Wright (6), Kawaguti and Mabuchi
(35), Salvini-Plawen and Mayr (55).

ocelli in Turbellaria, primarily on the basis of the presumed derivation
of the original Pericalymma larvae from the polyclad Lobophora larvae
(51,53). Because the original annelid organization was adapted to a
burrowing habit (11,12), and because the organizational level of Polychaeta must be regarded as successive to that of Oligochaeta (11,57),
the transfer of turbellarian photoreceptors can be comprehended only if
it occurred via the larvae. Because of a different photoreceptor
structure in the Lobophora (49) and sipunculidan-polychaetan larvae
(purely microvillar), their homology can be argued only under the
premises discussed above (see Notoplana). Additional doubts come from
the fact that in Echiurida only the worm-like larvae of Bonneliidae show

ocelli; neither the Trochophorae of other Echiurida nor the free-living
larvae (Lasiphora, Agasoma) of Myzostomida possess photoreceptors (see
53). Why should they have been reduced if once present?

There is absolutely no possibility of homologizing the converse
cerebral photoreceptors of rhabdomeric structure in Sipunculida or
Polychaeta (as well as Arthropoda s.1.) with the above inverse, poster-
ior-cerebral pigment-cup ocelli in these groups (lines 31 and 34 in
55, p.218). They are developmentally and morphologically two different
elaborations that supersede and replace each other (see 1, 2, 10).
The presumably early inverse pigment-cup ocelli are present in larvae
in general (1, 2, 20, 31) as well as in adults of limicoline or
tubicolous polychaetan groups (see 22, 55) and in most Archiannelida
(including the proximal receptor cell in Nerilla, 19). The converse
organs derived from epidermal cells are typical replacements for new
functional requirements as in the Pelago-Benthosphaera (Sipunculida;
53) and in almost all carnivorous Phyllodocemorpha and Eunicemorpha
(see also 20); an analogous supplementation occurred with the distal
receptor cell in Nerilla (Archiannelida, 19).

With respect to the presumed evolutionary pathway of Annelida (Polychaeta) and of Sipunculida, the converse cerebral photoreceptors of
each group also probably cannot be homologized (Fig. 4): primitive
coelomates originated as a result of burrowing movements in sediment,
and annelids subsequently adapted to this habit better by means of
coelomic segmentation (see 11). Such continuity of infaunal habit in
annelids (see Oligochaeta) made additional photoreceptors superfluous
until an epibenthic manner of living was preferred. Such a condition
renders it impossible to equate the converse photoreceptors of the
nonsegmented Sipunculida with the converse eyes of Polychaeta (see
Fig. 4). With regard to the Schizocoelia (Arthropoda s.1.), a common
epibenthic population of Oligochaeta may already have acquired converse
photoreceptors before giving rise to Polychaeta and Schizocoelia
respectively (see Fig. 4).

As demonstrated by their distinct developmental patterns and their
suitability for different habitats, inverse pigment-cup ocelli and converse photoreceptors are two distinctly different cerebral eyes adapted
to different photosensitive requirements. Despite their similar structural elaboration of microvillar enlargements, they differ functionally
as well as morphogenetically and no mutual homology can be assigned to
them. The ultrastructural pattern also suggests that the pigment-cup
ocelli belong to the ganglionic diverticular type (no rudiment of ciliary
structure has yet been found even in larvae), whereas the everse eyes
doubtlessly were derived from the epidermis (rhabdomeric) as confirmed
by the occasional occurrence of ciliary structures. However, the early
developmental patterns of converse photoreceptors has not been investigated in the larvae (Trochophora, etc.), and we may therefore point to
the evidence by Holborow (31, p.243) in a carnivorous Phyllodocemorpha
"that an eyelike structure of ciliary origin is in the progress of development in the larva." If the formations with whorled membranes derived from multiple cilia become the adult photoreceptors (with replacement of receptoral cells), it is not necessary to provide additional
evidence to rebut Eakin's postulation.

Mollusca

The main emphasis concerning a common line of rhabdomeric photoreceptors is placed on the sense organs in Polychaeta and Mollusca (18, 48). There is, however, evidence that molluscan photoreceptors are independent acquisitions only within the phylum. Cerebral photoreceptors are elaborated in Gastropoda, Siphonopoda (cephalopods), and Bivalvia. In the former two classes the organs of rhabdomeric ultrastructure can be regarded as homologous only owing to their probable common origin in more advanced Tryblidiida that were already provided with a distinctly differentiated head region no longer covered by a mantle with shell (see 54). The homology of these eyes with cerebrally innervated photoreceptors at the first branchial filament in Bivalvia-Pteriomorpha (= Filibranchia plus Pseudolamellibranchia) and their larvae can only be argued with reservation (55). With respect to their evolutionary condition in larvae (see below) and to a head region continously covered by a shell during evolution, the bivalve ocelli have evolved as independent organs. This is supported by the fact that the primitive, i.e., protobranch, bivalves have no photoreceptor formation at all that corresponds to their fairly conservative evolutionary organization. If the Bivalvia originated on soft, i.e., deeper bottoms, rather than in shallow photic regions (compare Protobranchia), the invasion into littoral zones only occurred as a later event by means of paedomorphous byssus attachment (58, 64). Only this new environment led to the respective adaptation of autobranch bivalves (most conservatively retained with the Pteriomorpha) including the differentiation of protonephridia and of eye spots (52, 54), as supported by their behavioral significance (see 47). Moreover, ultrastructural analysis reveals that the receptor cells still represent a fairly unmodified mixed type condition (microvilli plus an intact cilium) that suggests a more recent acquisition.

The argument of Rosen et al. (48) that the absence of cerebral eyes in five of the recent classes is due to their (secondary) living in aphotic environments is valid only for Caudofoveata, for Scaphopoda, and probably for Tryblidiida (compare, however, *Vema* hyalina, 41) or also for protobranch Bivalvia (see above). Neither the Placophora (typically adapted to the tidal littoral) nor the Solenogastres (freemoving epibenthic predators associated with Cnidaria) live in especially aphotic habitats. Owing to a homogeneous dorsal (mantle) cover of a chitinous cuticle with aragonitic scales and/or seven to eight shell plates or of a shell (Tryblidiida, Bivalvia, Scaphopoda) no cerebral photoreceptors were differentiated prior to the restriction of the mantle (shell) to the pallio-visceral body, a development correlated with the elaboration of a free head region (54, also 27).

On the other hand, the original larval type, the Pericalymma, possesses no trace of photoreceptors in either Solenogastres or protobranch Bivalvia. Also the transitional larval type (Stenocalymma) in Scaphopoda and Solenogastres, as well as all Pseudotrochoporae of Solenogastres and of Gastropods (archaeogastropod and other) are devoid of photoreceptors (50). The eyes in gastropod Veliger and bivalve Pediveliger/Veliconcha, however, are typically accelerated adult organs correlated with an extended presence of these advanced larvae as holopelagic-planktotrophic organisms (53). True larval photoreceptors have originated in the Pseudotrochophora of Placophora and may have differentiated in connection

FIG. 4. Relationship of Mollusca and Coelomata with the main evolutionary events indicated and supplemented by the origins of photoreceptors (see text).

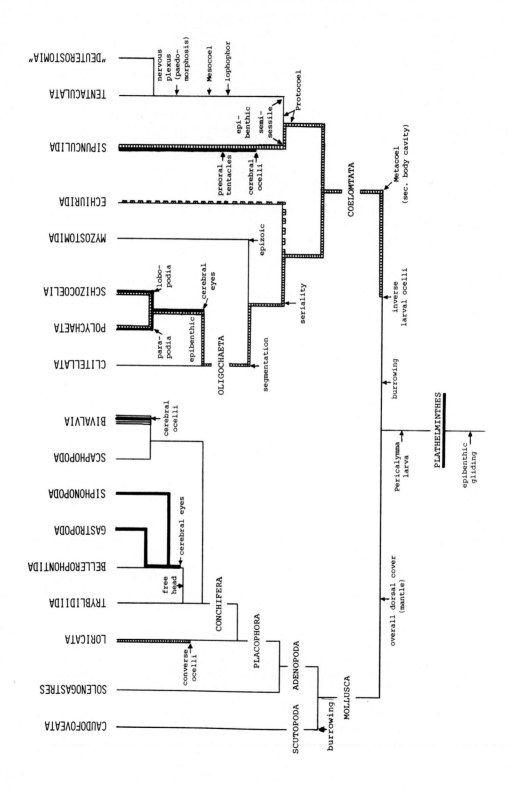

with their settlement, so that they may persist even for a period after
metamorphosis. Owing to their lack of photosensitive significance in
the adult, however, these ocelli are replaced by the photoreceptor cells
of the aesthetes (see 23, 24, 28). In addition to such ontogenetic
differences between the larval ocelli in Placophora-Loricata and the
adult cephalic eyes in Gastropoda and Siphonopoda or Bivalvia, and to
the eyeless state of all fairly primitive larvae in other molluscs, also
the innervation of the placophoran ocelli by the lateral nerve cord
(similar to the aesthetes with their photoreceptors) renders no homology
possible. The opposing speculation by Rosen et al. (48) does not include factual evidence of homology (see 46): the cerebral ocelli of
other larvae are inverse organs and originate from or close to the
apical sensory plate to which the cerebral ganglion is closely associated.
Even if the ocelli are secondarily shifted into a postoral (posttrochal)
position as in bivalves, the innervation remains cerebral (see 43). In
contrast, the lateral ocelli in Placophora are converse and originate
epidermally in loco, namely in a postoral (posttrochal) position, from
the later medioventral mantle-epithelium close to the mantle edge. In
addition, the medullary cords in Placophora (and Solenogastres, Turbellaria, Nemertini) develop as outgrowths from the cerebral center, only
secondarily entering into association with the ocelli (compare 36 with
29). Placophoran ocelli are ultrastructurally rhabdomeric in type but
include one or two supplementary cilia in which the membrane "is somewhat irregular and has a few short microvilli-like projections" (24,
p.54). Hence this organ also represents a mixed type of receptors and
suggests a fairly new acquisition.

There is no factually supported argument in favor of a hereditary
transmission of molluscan photoreceptors, either adult eyes or larval
ocelli, from their non-molluscan ancestors. In contrast, all available
comparative-anatomical, ontogenetic, and evolutionary-functional data
distinctly point to an independent acquisition of cephalic eyes in adult
forerunners of Gastropoda and Siphonopoda (cephalopods), and to a
secondary differentiation of lateral ocelli in placophoran larvae
successive to the original Pericalymma (Fig. 4). The lateral photoreceptors in Placophora are not comparable to the inverse pigment-cup
ocelli or accelerated converse eyes of the apical sensory plate in
Sipunculida and Polychaeta: the alleged photoreceptors in the common
adult ancestor of (adult) molluscs and (adult) coelomates (see Fig. 4)
should then be converse cerebral eyes; however, these are not existent
in platyhelminths and the theory is not consistent with a synorganized
differentiation in molluscan evolution (mantle cover, etc. see 27, 52,
54, 58).

The similarities between converse cerebral eyes of molluscs and
annelids, because of their microvillar receptor structure (18, 48) can
thus be considered as stemming from a similar functional requirement.
The condition in Pterotrachea (14) is an exception. Be it a positive
phototaxis (increase of intensity) or a registration of the existence of
light (see 7), sensitive epidermal cells, provided with some microvilli
plus one or more cilia, are convergently induced to elaborate the
respectively more appropriate microvilli. Such a condition has nothing
to do with homology or an endogenous, hereditary canalization. Only
other comparative characters (see above) can support or demonstrate a
possible homology. They do not do so, however, in cerebral photoreceptors of molluscs and sipunculids or polychaetes.

5. A POLYPHYLETIC THEORY

All data advanced so far in this report demonstrate a polyphyletic elaboration of diffusely light sensitive cells that have sometimes, analogously, become fairly similar organs (see 55). Such convergence becomes particularly obvious when the "limitation of organic design" (46) reflects similar or even identical requirements as in the case of microvillar photoreceptive organelles. It may indeed be a mere academic question (18) whether we typify two different microvillar receptors from their origins as epidermal-rhabdomeric and as ganglionic, aciliate-diverticular type. We think, however, that such a precise distinction may contribute to our understanding of photoreceptors as e.g.,in Salpa, especially because there exist neural/ganglionic receptors that show no pleats at all (see 3).

As demonstrated in most groups by their "exceptions", the theory of two evolutionary lines of photoreceptive structure in Bilateria, ciliary versus rhabdomeric (15, 16, 18), cannot be maintained. Within that frame it is not of essential significance whether there are 40 to 65 or more (or even less) separate phyletic lines (see 55). The major question is: are they mon-, di-, (tri-) or polyphyletic? And there is a plain answer: polyphyletic! Even in Eakin's revised concept (18) there is a clear acknowledgment of the polyphyletic condition, including the postulate that the microvillar structure in cerebral photoreceptors of Protostomia constitutes just one of those polyphyletic lines. Eakin's revised concept would only apply to the taxon Spiralia (such as the line of Ascidiacea-Vertebrata). Yet even that concept is too perforated to represent an acceptable statement. The same criteria that serve to reveal the convergences in protostomes and echinoderms for example, also help to elucidate analogies within the Protostomia themselves, namely, the criteria of the homology theorem based upon comparative anatomy, ontogeny, and functionally synorganized conditions. Apart from the "exceptions" to microvillar cerebral photoreceptors in Lobophora larvae (Polycladida), in Tholophora larvae (Kamptozoa), in Gastrotricha, in Heteropoda (Gastropoda), in Bugula larvae (Bryozoa), and in probable additional lines (see 9, 14, 21, 31, 61), those criteria also indicate independent acquisitions of photoreceptors in Placophora and in Conchifera as well as a diphyletic status of ocelli and eyes in Sipunculida and Polychaeta (Fig. 4).

There are still large gaps in our knowledge of morphological (ultrastructural) conditions. The findings in larval and adult polyclad Turbellaria, exemplify our need to learn about the morphogenesis of the photoreceptors. We may ask, for example, whether the bivalve larvae in its photosensitive phase (the eyed Veliconcha, see 7) recognizes a light stimulus by ciliary receptors of the mixed organ, prior to turning to microvillar reception in the photonegative phase? A similar condition appears to exist in Polycladida and other lines such as Placophora or different mantle eyes of Cardiacea-Pectenacea (Bivalvia) where a photoreceptor may be abandoned or replaced in predominance by another type. Such change is indirectly supported by the condition in insect photoreceptors where epidermal cells of extraretinal origin are recruited into the photoreceptor during growth (see 26, 33).

More information about the ontogenetic differentiation of photoreceptors may add to our understanding of congenital ultrastructure (homology) and functional correlation (analogy). A homologous condition might then be restricted to a special elaboration of the receptor structure, e.g., "ciliary type with disks" (vertebrates); with respect to the "limitation of organic design" (46) such specification will become a very difficult task, however, in regard to different subtypes of microvillar structure. Accordingly, we are now as before left with a strict homology theorem (see above); the ultrastructural appearance can only serve to confirm or to point to a possible homology within an already well-established relationship at a lower supraspecific level (see 55, 62).

SUMMARY

1. Both a mutation theory (Eakin) and an induction theory (Vanfleteren and Coomans) coincide in principle with the concept of polyphyletic origin of photoreceptors (Salvini-Plawen and Mayr).
2. The polyphyletic diversity of photoreceptors appears to be due to different functional requirements at different times during radiation of evolutionary lines.
3. Structural diversity appears to be derived from diffuse light sensitive organs provided with both cilia and microvilli.
4. Polyphyletic elaboration of photoreceptors resulted in four different structural types: Ia, unmodified cells with (unpleated) cilia and microvilli (mixed types); Ib, cells with modified cilia (ciliary types); Ic, microvillar cells with vestigial cilia (rhabdomeric types); II, cells that have become aciliate before elaboration (ganglionic diverticular types).
5. Comparative analysis of photoreceptors during ontogenesis (Turbellaria-Polycladida) and phylogenesis (Bivalvia) confirms the hypothesis that the structural organization of the photosensitive organs can be changed according to new functional requirements (replacement or modification of existing cells).
6. Ultrastructural analysis of photoreceptors cannot serve as the only basis for determining major phylogenetic relationships; it may serve, however, to confirm or to elucidate a possible homology within monophyletic lines of already well-established relationships.
7. The polyphyletic status of microvillar (rhabdomeric/diverticular) cerebral photoreceptors (Protostomia) is confirmed.
8. A detailed correlation between structural types and functional requirements of photoreceptors, especially with regard to changes during life cycles, has yet to be fully elucidated.

REFERENCES

1. Åkesson,B.(1958): Undersökn. Öresund, 38:1-249.

2. Åkesson,B.(1961): Galathea Rep., 5:7-17.

3. Autrum,H.(1979): In:Handbook of Sensory Physiology, Vol. VII/6A, edited by H. Autrum, pp.1-22. Springer-Verlag, Berlin, Heidelberg, New York.

4. Barber,V.,Evans,E., and Land,E.,(1967): Z. Zellforsch. Mikrosk. Anat., 76:295-312.

5. Barber,V, and Land,M.(1967): Experientia, 23:677-678.

6.   Barber,V.C., and Wright,D.E.(1969): J. Ultrastruct. Res., 26:515-528.

7.   Bayne,B.(1964): Oikos, 15:162-174.

8.   Brandenburger,J.L.,Woollacott,R.M., and Eakin,R.M.(1973):
     Z. Zellforsch. Mikrosk. Anat., 142:89-102.

9.   Burr,A.H., and Burr,C.(1975): J. Ultrastruct. Res., 51:1-15.

10. Clark,R.(1956): J. Exp. Biol., 33:461-477.

11. Clark,R.(1964): Dynamics in Metazoan Evolution: The Origin of the
     Coelom and Segments, 313pp. Clarendon Press, Oxford.

12. Clark,R.(1979): In: The Origin of Major Invertebrate Groups, edited
     by M. House, pp.55-102. Academic Press, London.

13. Crisp,M.(1972): J. Mar. Biol. Ass. U.K., 52:437-442.

14. Dilly,P.N.(1969): Z. Zellforsch. Mikrosk. Anat., 99:420-429.

15. Eakin,R.M.(1963): In: General Physiology of Cell Specialization,
     edited by D. Mazia and A. Tyler, pp. 393-425. McGraw-Hill, New York.

16. Eakin,R.M.(1965): Cold Spring Harbor Symp. Quant. Biol., 30:363-370.

17. Eakin,R.M.(1968): In: Evolutionary Biology, Vol. 2, edited by
     T.Dobzhansky, M.K.Hecht and W.C.Steere, pp. 194-242, Appleton-
     Century-Crofts, New York.

18. Eakin,R.M.(1979): Am. Zool. 19:647-653.

19. Eakin,R.M.,Martin,G.C., and Reed,C.T.(1977): Zoomorphologie, 88:1-18.

20. Eakin,R.M., and Westfall,J.A.(1964): Z. Zellforsch. Mikrosk. Anat.,
     62:310-332.

21. Ehlers,B., and Ehlers,U.(1977): Zoomorphologie, 87:65-72.

22. Ermak,T., and Eakin,R.M.(1976): J. Ultrastruct. Res., 54:243-260.

23. Fischer,F.(1978): Spixiana, 1:209-213.

24. Fischer,F.(1980): Spixiana, 3:53-57.

25. Gorman,A.L.F.,McReynolds,J.S., and Barnes,S.N.(1971): Science,
     172:1052-1054.

26. Green,S., and Lawrence,P.(1975): Wilhelm Roux' Arch., 117:61-65.

27. Haas,W.(1980): Haliotis, 10:66 (abstract) = Malacologia (in press).

28. Haas,W., and Kriesten,K.(1978): Zoomorphologie, 90:253-268.

29. Hammarsten,O., and Runnström,J.(1926): Zool. Jahrb. Anat., 47:261-318.

30. Hermans,C.O.(1969): Z. Zellforsch. Mikrosk. Anat., 96:361-371.

31. Holborow,P.(1971): In: Fourth Eur. Mar. Biol. Symp., edited by D.J. Crisp, pp.237-246. Cambridge Univ. Press, Cambridge.

32. Hughes,H.P.I.(1970): Z. Zellforsch. Mikrosk. Anat., 106:79-98.

33. Hyde,C.(1972): J. Embryol. Exp. Morphol., 27:367-379.

34. Hyman,L.(1951): The Invertebrates, Vol. 2, 550 pp. McGraw-Hill, New York, Toronto, London.

35. Kawaguti,S., and Mabuchi,K.(1969): Biol. J. Okayama Univ., 15:87-100.

36. Kowalevsky,M.A.(1883): Ann. Mus. Hist. Nat. Marseille, Zool. 1:1-46.

37. Kuhn,O.(1960): Stud. gen., 13:477-491.

38. Lang,A.(1884): Fauna Flora Golf Neapel, Monogr. XI, 688 pp. W. Englemann, Leipzig.

39. Laverack,M.(1968): Oceanogr.Mar. Biol. Ann. Rev., 6:249-324.

40. MacRae,E.K.(1966): Z. Zellforsch. Mikrosk. Anat., 75:469-484.

41. McLean,J.(1979): Contrib. Sci. Nat. Hist. Mus. Los Angeles County, 307:1-19.

42. Millott,N.(1968): Symp. Zool. Soc. London, 23:1-36.

43. Pelseneer,P.(1900): Arch. Biol. (Liege), 16:97-103.

44. Plate,L.(1924): Allgemeine Zoologie und Abstammungslehre. II. Die Sinnesorgane der Tiere, 806pp. Fischer-Verlag, Jena.

45. Reisinger,E.(1972): Z. Zool. Syst. Evolutionsforsch., 10:1-43.

46. Rieger,R., and Tyler,S.(1979): Am. Zool., 19:655-664.

47. Rosen,M.,Stasek,Ch., and Hermans,C.(1978): The Veliger, 21:10-18.

48. Rosen,M.,Stasek,Ch., and Hermans,C.(1979): The Veliger, 22:173-178.

49. Ruppert,E.E.(1978): In: Settlement and Metamorphosis and Marine Invertebrate Larvae, edited by F.S.Chia and M.E.Rice, pp. 65-81, Elsevier/North Holland Biomedical Press, New York.

50. Salvini-Plawen,L.v.(1969): Malacologia, 9:191-216.

51. Salvini-Plawen,L.v.(1972): Z. wiss. Zool., 184:205-394.

52. Salvini-Plawen,L.v.(1980a): Malacologia, 19:249-278.

53. Salvini-Plawen,L.v.(1980b): Zool. Jahrb. Anat., 103:389-423.

54. Salvini-Plawen,L.v.(1981): Atti Accad. naz. Lincei (Roma), 11:1-59.

55. Salvini-Plawen,L.v. and Mayr,E.(1977): In: Evolutionary Biology, Vol. 10, edited by M.K.Hecht, W.C.Steere and B.Wallace, pp.207-263. Plenum, New York.

56. Salvini-Plawen,L.v., and Splechtna,H.(1979): Z. Zool. Syst. Evolutionsforsch., 17:10-30.

57. Siewing,R.(1978): Zool. Jahrb. Anat., 99:93-98.

58. Stanley,St.(1972): J. Paleontol, 46:165-212.

59. Stasek,Ch.(1966): Occas. Pap. Calif. Acad. Sci., 58:1-9.

60. Storch,V.(1979): Am. Zool., 19:637-645.

61. Teuchert,G.(1976): Zoomorphologie, 83:193-207.

62. Vanfleteren,J.R., and Coomans,A.(1976): Z. Zool. Syst. Evolutionsforsch., 14:157-169.

63. Wilkens,L., and Larimer,J.(1976): J. Comp. Physiol., 106:69-75.

64. Yonge,C.M.(1962): J. Mar. Biol. Ass. U.K., 42:113-125.

65. Yoshida,M.(1979): In: Handbook Sensory Physiology, VII/6A, edited by H. Autrum, pp.581-640. Springer-Verlag, Berlin, Heidelberg, New York.

Subject Index

Subject Index